Mason Patrick and the Fight for Air Service Independence

MASON PATRICK

AND THE FIGHT FOR AIR SERVICE INDEPENDENCE

ROBERT P. WHITE

SMITHSONIAN INSTITUTION PRESS
Washington and London

©2001 by Robert P. White
All rights reserved

Copy editor: Therese D. Boyd
Production editor: Robert A. Poarch
Designer: Janice Wheeler

Library of Congress Cataloging-in-Publication Data
White, Robert P.
    Mason Patrick and the fight for air service independence / Robert P. White.
        p. cm.
    Includes bibliographical references and index.
    ISBN 1-56098-943-2 (alk. paper)
    1. Aeronautics, Military—United States—History—20th century.
    2. Patrick, Mason M. (Mason Mathews), 1863–1942. I. Title.
UG633.W395 2001
358.4′00973′09041—dc21   2001020403

British Library Cataloguing-in-Publication Data is available

Manufactured in the United States of America

08 07 06 05 04 03 02 01   5 4 3 2 1

∞ The paper used in this publication meets the minimum requirements of the American National Standard for Information Sciences—Permanence of Paper for Printed Library Materials ANSI Z39.48-1984.

For permission to reproduce illustrations appearing in this book, please correspond directly with the owners of the works, as listed in the individual captions. The Smithsonian Institution Press does not retain reproduction rights for these illustrations individually or maintain a file of addresses for photo sources.

To my wife, Mary, and our children

Contents

Acknowledgments **ix**

Introduction **1**

1. From the Wright Brothers to World War I **8**
2. Army Engineer from the Hills of West Virginia **15**
3. Patrick Wins His War within a War **26**
4. Patrick and the Postwar AEF Air Service, 1918–1919 **36**
5. The First Round in the Postwar Fight for Air Service Independence, 1919–1921 **44**
6. "To Command in Fact as Well as in Name" **58**
7. The Lassiter Board and "Fundamental Conceptions": Spadework for the 1926 Air Corps Act **72**
8. Patrick Takes on a Pair of Wings, the Navy, and the Army General Service Schools **85**
9. Patrick's Search for Economy and Efficiency **94**
10. The Fallout from the Lassiter Report and the Fall of Billy Mitchell **110**
11. The Air Corps Act and Its Aftermath **122**

12. Conclusion **132**

   Notes **139**
   Bibliography **175**
   Index **183**

Acknowledgments

The cumulative expertise of dozens of individuals made this book possible. I only strung their pearls of wisdom together. Professor Allan R. Millett patiently contributed to the string of thought and certainly had his work cut out for him, as whatever good turn of phrase is within, it is due in great part to him. I am also indebted to Bruce Bingle, who as one of Dr. Millett's students in the early 1980s took the first serious look at Mason Patrick's contribution to the Air Service. He broke new ground and eased my search to expand on Patrick's story.

While the search was made difficult by the fact that there exists no dedicated collection of Patrick papers, there were many who helped me find precious nuggets of information. Without them nothing could have been accomplished. I would like to thank the staff at the Air Force Historical Research Agency, Maxwell Air Force Base, especially Joe Caver, Archie Difante, Lynn Gamma, and Warren Trest—their extensive knowledge and assistance saved me many hours of labor. Duane Reed at the U.S. Air Force Academy Library, Special Collections Division, is an exceptional Air Force history archival authority. Two individuals at the U.S. Military Academy also provided individual assistance beyond the call of duty: Alan Aimone and Suzanne Cristoff. At the National Archives John Taylor set me on the proper path to research in that wonderland of Hollinger boxes. While there, I was well tended to by Will Mahoney, Ed Reese, and Mitch Yockelson, all of whom helped me find my way through numerous Finding Aids and provided still more assistance though the thousands of linear feet in Record Group 18. At the Library of Congress, Special Collections Branch, the entire staff enlightened me regarding the contents of the Mitchell and Pershing collections. I was made most welcome at the U.S. Army Institute for Military History at Carlisle Barracks, Pennsylvania, where David Keogh, Col. Jim McClain, and Dr. Richard Sommers provided excellent support. Mike Walker at the Naval Historical Center, Washington, D.C., always took the time to check arcane facts for me. I am most appreciative of the very accommodating spirit and helpful assistance of the staff of the Public Record Office, London, England, given my limited time in that most impressive archive.

At the Air Force History Support Office in Washington, D.C., David Chenoweth, Yvonne Kinkaid, Dr. Roger Miller, Dr. George Watson, and Dr. Richard Wolf came to the rescue with, respectively, photo expertise, archival minutia, clarification, inspiration, and computer resuscitation. The Air Force Historian, Dr. Richard Hallion, provided unlimited support.

At the USAF Air University Col. Phil Meilinger, Ph.D., Dr. David Mets, and Dr. Jim Titus have all given me encouragement and advice. My brother James gave me the motivation to finish sooner than I would have, and my mother's prayers doubtless ensured that I would indeed, ultimately see this in print. Special appreciation must go to Dr. Dan Mortensen, who read and reread. His many suggestions, encouragement, and unflagging friendship are all greatly appreciated.

Introduction

In the summer of 1923, in a cloudless sky above Bolling Field just outside of Washington, D.C., a student pilot of the Army Air Service soloed for the first time. When he lifted off the grass airstrip he began a series of acrobatic maneuvers that would culminate in his successful proficiency flight to become the oldest Junior Military Aviator in the history of the Air Service and the oldest rated officer ever in what would later become the United States Air Force. Major Gen. Mason M. Patrick, chief of the Army Air Service since October 1921, earned his wings that day at age fifty-nine. In December 1927, as General Patrick was about to retire at the mandatory age of sixty-four, he flew over Bolling Field again, but this time as chief of the vastly improved Army Air Corps. From his first solo flight to his last active duty sortie, Mason Patrick was responsible for six years of extraordinary change within the Army Air Service and its successor, the Army Air Corps.

Unfortunately, little is remembered of Patrick, who, in retrospect, was responsible for saving a fledgling air force from a variety of self-inflicted wounds and many competing and self-serving outside interests. In addition, the interwar period, and especially the decade following World War I, has attracted little research on the Air Service and Air Corps. In a popular and academic literary sleight of hand, it seems that if one knows the story of William "Billy" Mitchell, one knows everything. In the popular mind Mitchell and his travails personify and dominate the era. Mitchell, however, was only part of the story, a story that began with relative disinterest in the promise of flight on the part of American military planners—at least in comparison to the European powers of the day.

To understand the enormous challenge that ultimately faced Mason Patrick as head of the Air Service in the 1920s, one must appreciate the early character of American military aviation. For one, the orthodox character of the War Department fostered an innate lack of appreciation for new technology—in this case using the airplane as a weapon. This, combined with the government's ultraconservative fiscal policy and America's inherent isolationism, severely dampened any enthusiasm for military aviation until the nation's entry into World War I.

In Europe, on the contrary, there was enthusiastic support for aviation. The dawning of the air age on the continent contributed to a new culture of time and space that readily acknowledged the major impact of air technology on society as a whole. By comparison, the appreciation of air power, both military and civilian, was initially almost wholly lost on the American psyche. The modest advances that did occur in American aviation circles were due to a handful of dedicated entrepreneurs and scientists whom one historian termed the "invisible establishment."[1]

Exactly how much the American and European military establishments spent on aviation provides a more concrete comparison: between 1908 and 1913 the United States spent approximately $435,000 on Army and Navy aviation programs. By contrast, France alone spent $22 million during the same period. This is not to say that prior to World War I the United States made no attempts to plan for the military aviation needs of the future. Quite the contrary. There was a lot of talk, primarily in Congress, but little action.

It took an acutely embarrassing aviation performance during the 1916 Punitive Expedition into Mexico and America's entry into World War I to eliminate this lethargy. American aviation leaders had much to learn from their European counterparts. Such was the sad state of affairs during World War I that U.S. pilots, for the most part, flew second-hand European aircraft during the war and employed European air doctrine, there being no such thing as an indigenously developed American military aviation doctrine prior to or for several years after the war.

In revolutionary terms World War I was to American military aviation what the Spanish-American War was to the American Army in general. The need for dramatic reappraisal and action was obvious. This long-overdue reassessment led to many contentious confrontations between not only soldiers and airmen but also among the airmen. These disputes would not have been so disruptive if not for the fact that they were taking place in the midst of a war. Gen. John J. Pershing, as the American Expeditionary Force (AEF) commander, was caught up in these distractions. He also had to contend with the inflated egos of Mitchell and Benjamin "Benny" Foulois, the two airmen who vied for dominance in the formative months of the AEF Air Service. These problems ultimately led to the appointment of Pershing's good friend and West Point classmate, Major Gen. Mason Patrick, as chief of the Air Service, AEF.

Pershing fully realized the capabilities and disabilities of his two top Air Service officers: Mitchell was dogmatic, flamboyant, and an excellent combat commander; Foulois was the best "home-grown" senior officer the Air Service had produced up to that time, and he was a superior pilot. But neither Mitchell nor Foulois was a good administrator. Pershing put the AEF Air Service leadership problem into perspective when noting that they were "good men running around in circles." Pershing's solution was obvious: he appointed one of the strongest administrators he knew to get the Air Service to fly in single formation.

Mason Mathews Patrick graduated second in his 1886 West Point class. Due to his high class standing he was allowed to choose his career field and thus chose to be an engineer, a job he performed with drive and intuitive administrative skill up to the moment Pershing asked him to take over the AEF Air Service. Patrick's no-nonsense approach brought order to the personality-induced chaos that had engulfed the Air Service. That Pershing had to go outside of the Air Service to find a capable commander points up a major shortfall that would continue to plague the young air arm: lack of capable senior leadership. As General Patrick later noted in his diary, Pershing could very well have dismissed both Mitchell and Foulois if experienced air leadership had been waiting in the wings. With Patrick in charge the AEF Air Service began to provide the support that Pershing desperately required, but there were still problems with the way Mitchell and many of his contemporaries viewed their ultimate utilization as a combat arm. The seeds of strategic bombing doctrine that took root during the latter part of World War I, and the unanswered question of the airplane's full combat potential, would pit airman against soldier for the next forty years. Unresolved at the close of the war, these two issues would create divisive conflict in the immediate years to come.

At the Armistice on 11 November 1918 the Air Service had almost 200,000 personnel, 11,000 planes (of the 27,000 ordered) in forty-five aero squadrons, and forty-eight airfields, complemented by nineteen supply depots around the country.[2] The acquisition, training, and supply pipelines of the Air Service were running at peak capacity on Armistice Day, and late that same afternoon the Air Service began to demobilize. Unfortunately, there was little forethought concerning how demobilization would be accomplished, and even less consideration regarding the postwar Air Service.

At the close of the war Patrick remained in Paris to assist Pershing as an advisor to the American peace delegation. Another member of the West Point class of '86, Maj. Gen. Charles T. Menoher, a straight-laced artillery officer who had commanded the Rainbow Division on the Western Front, was made Air Service chief, the job Billy Mitchell coveted. The inevitable clash of wills ultimately resulted in Menoher's request to be relieved as Air Service chief and be returned to the Army of the line. When Patrick replaced Menoher and took over the Air Service in October 1921, again at the request of Pershing, the Army Air Service was in its death throes. The Air Service at the close of 1921, with about 10,000 personnel, was a mere skeleton compared to the aviation component of the AEF of World War I, and it was embroiled in doctrinal disagreements, severe fiscal deficiencies, and personal antagonisms. During World War I the AEF Air Service's difficulties of competent command centered primarily on personality problems. Such was the case once again, but additional issues such as the evolution of aircraft technology, new air-war fighting tactics, and a dearth of funding exacerbated tensions within the Air Service and its relationship with the War Department.

When Mitchell returned to the United States from Europe in March 1919 he was possessed of a visionary blueprint for a new military policy based on the omnipotence of air power. But Mitchell made some wrong assumptions, primary of which was that everyone would fully appreciate and readily implement his vision. Mitchell also manifested two traits, sensationalism and an uncompromising character. Menoher and Mitchell were in a dogfight from the start, and Menoher's brief tenure as chief of the Air Service allowed for little accommodation and compromise in the debate about what to do with both civil and military aeronautics. Mitchell won the confrontation handily. Menoher, stymied in his attempt to rein in Mitchell, was replaced by the strong, steady, and knowledgeable hand of Mason Patrick.

Initially, keeping the Air Service breathing, let alone attempting to gain autonomy, was a massive and problematic undertaking. Patrick realized both the advantages and disadvantages of Mitchell's talents and employed them accordingly. From October 1921 through 1927 it fell to Patrick to orchestrate the behind-the-scenes policies and politics that eventually resulted in the creation of the U.S. Army Air Corps in July 1926, along with an impressive five-year procurement program. In the long term Mason Patrick did more than anyone during this period to ensure the successful independence of the Air Force. The Army Air Corps, in essence, came into existence in spite of Mitchell, not because of him.

Patrick was as "low profile" as Mitchell was both "high profile" and "highbrow," but Patrick's aversion to sensationalist headlines only enhanced his effectiveness as an Army insider and an Air Service/Air Corps advocate. Ironically, Patrick agreed with much of what Mitchell espoused and voiced many of those same opinions in his congressional testimony, speeches, and endorsement of doctrine. Patrick was practical enough to know that the Air Service's survival depended on a doctrine that explicitly supported the need for autonomy. Most important, though, it was the degree of autonomy Patrick pursued that made him different. He was much more effective at enhancing the credibility of the Air Service than Mitchell, even though it was Mitchell who grabbed the headlines.

Ultimately, Mitchell resigned from the service in February 1926 after being found guilty of insubordination at the conclusion of a two-month court-martial, which was precipitated by Mitchell's 5 September 1925 broadside criticizing the "incompetency, criminal negligence and almost treasonable administration of the national defense by the war and navy departments."[3] Mitchell's court-martial ensured his martyrdom and enshrined his memory in Air Force history to the exclusion of other notable airpower advocates of the time.

Mason Patrick, Mitchell's boss, was so forgotten. It was Patrick who assumed the stewardship of the Army Air Service at two critical times and held in check the forces swirling in and about the Army Air Service. Mitchell contributed significantly to this conflict and paid a heavy price. Beginning in 1919 Mitchell had waged a campaign for Air Service independence and the acceptance of the airplane as the future (and perhaps current) arbiter of armed conflict. His prophecies took root in a number of true believers among Air Service officers.[4] Patrick was faced with controlling Mitchell's sensationalist and divisive tactics. The forces that confronted Patrick could either be brought into harmonious (or at least grudging) balance or, if left untethered, could forever sunder the promise of an independent air force. Patrick's successful and masterful role in this struggle has vanished from popular knowledge, being overshadowed to a great extent by the Mitchell controversy and its subsequent notoriety.

Mitchell's martyrdom crystallized overnight into Air Force hero worship and doctrinal mythology. Propagated by his disciples, his gospel of omnipotent air power was carried forward with missionary zeal and survives to the present day. One could argue that Mitchell's sensationalist approach to the primacy of air power fit perfectly into his role not as a leader of the Air Service but as a publicist.

The short-lived nova of Mitchell's court-martial lent an aura of urgency and legitimacy to the Air Service's fight for acceptance and independence, but it was Patrick's political "horse sense," influence, and agenda that ensured a victory with the ultimate creation of the Air Corps. Many interests were vying for dominance of the airpower question during this period, but the one man who saw the Air Service and Air Corps through it successfully was the man in charge, Mason Patrick.

Mitchell was not the only challenge on the road to Air Service independence. Patrick also had to contend with the War Department heads and the General Staff, the Navy, Presidents Warren G. Harding and Calvin Coolidge, key congressional air activists of the era, industrialists, inventors, and especially the group of "young Turks" within the Air Service itself. Pershing charged Patrick with bringing order to an organization that took inordinate—and many claimed unjustified—pride in its unique capabilities as a combat arm. The "young Turks" of the Air Service were true believers in the "wondrous capabilities" of the airplane. They and Mitchell, their outspoken publicist, demanded what could not be given at that point in time:

independence coequal with the Army and Navy. It would fall to Patrick to instill discipline in the "upstart" Air Service and, accomplishing that, to establish the service's credibility within the Washington bureaucracy.

Patrick had few doctrinal allies on the inside to aid his quest, but as an insider himself he had the latitude to push certain concepts that would be greeted with derision if voiced by other airpower advocates. He had the freedom to push the Air Service and Air Corps toward a more autonomous status, given his stature and good standing within the War Department and Congress. Not the least bit obsequious, he was not averse to a good fight. Patrick's confrontations with congressional committees, some members of the War Department, the Navy Department, and some of his own officers demonstrate his intelligence, wit, determination, and charm.

Patrick may have been part of the War Department establishment but he was indeed ahead of his time. He initially envisioned an Air Corps and Army relationship that was analogous to the relationship the Marines and Navy enjoyed: separate services within the same department. Full autonomy would come in time by employing a gradualist approach. Patrick was convinced that the ultimate solution was a unified and separate air force, to include all aeronautical development under one responsible and directing head, within a national defense structure. He had a road map to get there, and its name was legislation.

To begin this process, Patrick drafted a 1924 proposal that ultimately led to the creation of the Air Corps in 1926. Granted, due to political pressures, he did not get all he asked for, but it was a major step toward recognizing the unique status of the air force. Mitchell thought little of Patrick's 1924 initiative, and when it was passed in 1926 he derided the results as quite inadequate because total independence was not achieved.

The 1920s "roared" in part due to the politically charged atmosphere surrounding military aviation, and there has never been a period since when the American populace was so enamored of the airplane. The Patrick era boils down to a raucous and tenacious struggle involving a new technology and its evolving doctrine that spawned new military capabilities. Doctrine and strategy bumped up against politically induced fiscal realities. Patrick not only fought the congressional battle but ultimately successfully challenged the orthodoxy of the War Department as well. The result was the development and institutionalization of new doctrine. The officer corps of the Air Service and Air Corps demanded dramatic change in operational doctrine and what logically followed: resource reallocation. This officer coterie developed a war-fighting doctrine that inherently internalized the rationale for service independence. What made this doctrinal and independence debate so interesting is that it was based to a great extent on unproved theories, and what many would say were futuristic fantasies. A twist in the debate concerned the disagreements between the respective services and the doctrinal debate within the Air Service itself. Those who wanted no discussion of divorce from the Army were pitted against pilots who advocated displacing the infantry, the "queen of battle," with an all-powerful independent air force. This debate was at the crux of autonomy and service budget rivalries. If the Air Service was largely, if not exclusively, tied by doctrine to the ground-force mission, there existed no rationale to support autonomy, and there would be no need for additional monies to support the infrastructure and mission of a separate service. General Patrick, though he appreciated strategic air doctrine, urged that ground-attack (close air support and interdiction to a lesser extent) aviation be greatly enhanced, and he was true to this belief throughout his tenure as chief of the Air Service/Corps. In fact, in Patrick's entire time as chief of America's air arm, he never advocated that air assets assigned to a ground-support role have any independent mission. Nor would those forces be commanded by any other than the tactical or theater commander.

While General Patrick was practical in his outlook, he was also a progressive visionary in his quest to obtain as much autonomy for the service as possible. His was a balanced and successful approach to the advocacy of air power. Unlike Mitchell, Patrick represented an era of planned evolutionary change accomplished within a milieu of competitive revolutionary theories and conservative regulational tradition. Up against immense odds, the Air Service, under Patrick's guidance, was put on the path to independence. While Mitchell's airpower obsession ultimately got the better of him, his approach almost got the better of the Air Service, were it not for Mason Patrick. He manifested an unerring sensibility in guiding the Air Service to a realistically achievable degree of autonomy.

1. From the Wright Brothers to World War I

Initial U.S. military aviation programs had a most difficult time getting off the ground. The War Department's credibility was seriously undermined in 1903 when its Board of Ordnance and Fortification allocated a grant of $50,000 to Dr. Samuel P. Langley for the rights to his Aerodrome flying machine. In October, and again in December, the prototype plummeted ignominiously into the Potomac River upon launch from its houseboat catapult.[1] The Washington press was there in force.

From that inauspicious beginning things improved very little for military aviation. Prior to World War I, there were very few aviators and even fewer planes. The smattering of aviation assets that did exist had no doctrine for their employment. The War Department simply did not consider the aeroplane a serious adjunct to the military mission. Compared to what other governments invested in their military air effort up through 1913, the United States came in thirteenth in the world rankings. At the outbreak of World War I, the U.S. air arsenal contained a grand total of seventeen operational aircraft, all of them obsolete, and only nineteen pilots who took their chances flying them. By comparison, the Direction de l'Aeronautique Militaire in France had over 600 modern military aircraft and Britain's Royal Flying Corps flew 168 state-of-the-art airplanes. Both of these organizations were, by the way, autonomous military arms.[2]

The United States trailed far behind all other industrialized nations in every other aspect of military aviation as well. The American mindset had yet to fully appreciate this new technology. A far more dynamic vision had taken hold in Europe where revolutionary changes had occurred prior to and at the turn of the century. Europeans were quite cognizant of new transportation technology that put them within easy reach of their neighboring states. Significantly, this new technology of "faster time" was not lost on the military staffs and inventive geniuses of Europe.[3]

But the Americans remained relatively aloof from these concerns. Granted, they had invented the aeroplane, but given their insularity from Europe and its political perturbations, not to mention a very tight-fisted attitude toward military budgets (less than $125,000 a year was devoted to the air branch from 1911 through 1914), there was no need to exploit such an innovation.[4] Granted, from the European security perspective, there were very good reasons to invest in this new technology, not the least of which was an ongoing arms race begun in the early 1900s in the midst of mutual hostility backed up by two major military alliances. The Atlantic Ocean and the admonition to refrain from entangling alliances certainly assisted the stagnation of military aviation in the United States during this period.

The War Department, taking a rather blasé attitude toward this upstart division of the Signal Corps, issued the first military aviators certificates, which were nothing more than a typewritten notice, signed by a captain in the adjutant general's office, that a notation had been entered in the officer's permanent record noting "military aviator" status.[5] Until 1912 Army pilots did not even qualify for flight under any written military standards. Instead, they qualified for an aviation license issued by the Fédération Aéronautique Internationale. It took a concerted effort on the part of the chief signal officer to obtain authorization for an appropriately designed military aviator badge, which was finally authorized in May 1913.[6]

In retrospect, one would assume that when World War I began in 1914 the U.S. War Department would have taken at least a passing interest in military aviation. Such was not the case. Fewer than 200 individuals worked in the entire Army aviation establishment. Only thirty aircraft had been acquired, the first of which was already hanging in the Smithsonian Museum, and there was but one military aviation school.[7]

When the Villista raid against Columbus, New Mexico, prompted the Punitive Expedition in 1916, the 1st Aero Squadron (the first tactical air unit to be put in the field) had only eleven qualified pilots assigned and eight obsolescent aircraft to perform much-needed reconnaissance.[8] When the United States entered World War I in 1917, fifty-five officers were assigned to the Aviation Section of the Signal Corps, with fewer than 300 aircraft.[9] Only twenty-six of those fifty-five officers were qualified pilots.[10] The United States was caught vastly unprepared for a war in the air when confronted by such a daunting challenge. General Patrick, when commenting on this state of affairs many years later, stated that when compared to European military aviation, the Army's Aviation Section/Air Service was "a negligible quantity."[11]

The progress of American military air power, between 1909, when the Army accepted its first airplane, and 1917, when America entered World War I, moved glacially forward. It remained viable only through the efforts of the energetic aviators who were addicted to this new machine that promised so much. The U.S. Congress, in its characteristically parsimonious fashion, authorized a direct appropriation for military aviation of only $125,000 in 1911, only half of the original proposal tacked onto the Army appropriations bill as a rider just as Congress was hurriedly preparing for an end-of-session adjournment.[12]

In retrospect, Congress cannot be held completely accountable for the abysmal state of military aviation prior to World War I. Given the fact that there was little attempt to professionalize the aviation officer corps, the War Department shares much of the blame. On balance, Congress contributed more to military aviation than the War Department did. In 1913 Congressman James Hay of Virginia, chairman of the House Military Affairs Committee, introduced a bill calling for creation of a separate aviation branch for the Army. The War Department opposed it, as did a certain Army captain by the name of Billy Mitchell. The bill failed to pass, but a 1914 version enacted into law on 18 July formally created an Aviation Section within the Signal Corps and authorized 60 officer and 160 enlisted billets. The seeming largess of these authorizations was actually minuscule: aeronautics accounted for less than 0.4 percent of Army personnel as a whole.[13] Significantly, another Army administrative requirement, the so-called Manchu Law, had a deleterious effect on Army fliers. This regulation restricted line officers to only four years of consecutive detached duty outside of the branch in which they were commissioned. Army pilots would no sooner become lever, stick, wheel-rudder, and wheel-yolk (types of aircraft-control systems) proficient and serve their geographic tour before they were reassigned to their primary duty.

Not only was there a lack of administrative regulations that protected Army aviation, there was also a dearth of standardization across the board. For example, the Army did not adopt a standard control system on military aircraft until 1917 when the "stick" system became the official norm.[14] Benjamin (Benny) Foulois, at that time the most experienced Army flier, did not write his "Provisional Aeroplane Regulations" and "Flying Safety Rules," the first guides in an effort to standardize aeroplane maintenance and flight techniques, until 1911.[15]

Up through the American entry into World War I, there was no effort to standardize or institute professional military aviation education. Military aviation doctrine was nonexistent. Perhaps more important, without any doctrine to guide them, not a single aviation unit had been trained for combat operations. Of course, training for combat would have been difficult, as the U.S. Army had no combat aircraft to train with. Army aviators had no combat pursuit plane experience and a war had been underway in Europe for almost three years.[16] Even with the ongoing precedent of pursuit combat in European skies, the U.S. Army aviation focus was on observation, air-ground cooperation, and artillery spotting. There were some bright spots: bombs and bombsights were invented and tested, and experiments were conducted with the highly effective Lewis machine gun, in addition to aerial photography and wireless communication. But there was little follow-up testing. Before World War I the aeroplane was viewed by the War Department as strictly a reconnaissance asset, and the General Staff was seemingly determined to keep it that way.[17] World War I, though, forced the hand of the General Staff, for the U.S. Army was about to witness a renaissance in American military aviation.

On 6 April 1917, when America intervened upon a stalemated field of battle, five U.S. Army Aviation Section aviators were in Europe: one in London as an assistant military attaché, three in French flying schools, and the fifth, Maj. William "Billy" Mitchell, an observer in Spain who had arrived on the continent less than one month before. Five weeks later Major Mitchell was Lieutenant Colonel Mitchell and involved in numerous aspects of the war effort. He actually took part in an infantry attack at the front, being the first American military member to come under fire. He was also the first American officer to fly over the front lines, accompanied by a French pilot, of course.[18]

From this beginning, the Air Service slowly evolved into a relatively dynamic and effective fighting force, but it was never a decisive factor of the overall war effort. The infantry of the respective belligerents were the decisive arbiters of battle. All in all, World War I military aviation was concerned primarily with reconnaissance in support of troops on the ground. But while aviation may not have been a decisive factor, it was not insignificant either: aerial photography, artillery spotting and adjustment, and infantry contact were all-important missions. Both Billy Mitchell (early on) and Mason Patrick (not much later) understood that to accomplish these missions, as well as the bombing and ground-attack missions, it was essential to control the air. It was only a short intellectual jump from total air control to the requirement for an independent air force. But an independent air force required, among other things, much improved tactics and technology. In time, though, significant advances took place in aviation technology and air-war tactics were refined.[19] Given the state of airplane technology at the beginning of hostilities, advances were definitely needed. In the realm of aircraft design, especially with regard to the pursuit mission, no sooner had a plane gone from drawing board to fabrication than it was obsolete.[20]

The U.S. Air Service, flying Allied hand-me-down aircraft, was always behind the technology curve. In the first year in Europe, the Air Service was crippled by a tumultuous turnover of personnel, heated arguments, and lingering antagonisms, many spawned by Billy Mitchell. At the outbreak of the war Mitchell found himself in charge of the U.S. air effort in Europe and he proceeded to bombard the War Department with hundreds of cables with a host of requests and observations. One must note that at this time the War Department was receiving, on average, 40,000 messages a day and blithely ignored many of Mitchell's missives.[21]

In the coming months, conflicting personalities and recommendations precipitated numerous conflicts within the Air Service organization. Mitchell's freewheeling days were pretty much at an end by the time General Pershing and his AEF staff arrived in June 1917. On the recommendation of the National Advisory Committee for Aeronautics (NACA), Signal Corps commander Gen. George O. Squier then sent the Bolling Commission to Europe to determine what types of aircraft the United States should build for the war effort. Ultimately, the ideal ratios of pursuit, observation, and bombardment aircraft requested by different organizations, committees, and individuals (the Bolling Commission, the War Department, the Signal Corps, General Pershing, Billy Mitchell) all differed significantly and led to heated debates. Thrown into this admixture were Brigadier General Foulois and a hundred-man staff chosen at Signal Corps headquarters in Washington, who arrived in Paris in November 1917 to take over the AEF air-war effort.

As if there were not enough problems associated with the Air Service, AEF, the arrival of the "Washington crowd" precipitated other difficulties, first among which was Mitchell's deep resentment of Benny Foulois. Although Mitchell did not say so publicly, in writing to his sister Ruth he conceded that Foulois "meant well and had some aviation experience."[22] But Mitchell knew full well of Foulois's extensive career in aviation; he had learned the rudiments of flight at College Park, Maryland, in 1909 under the tutelage of the Wright brothers[23] and then learned to fly on his own and solo the following year at San Antonio, Texas. It was as a first lieutenant that Foulois wrote the original "Provisional Aeroplane Regulations" for the Signal Corps.[24] In 1915 and 1916 he established the Signal Corps Aviation Center at San Antonio and then organized and commanded the first U.S. Army tactical air unit, the 1st Aero Squadron, and led it into Mexico in 1916. In March 1917 Foulois was called upon by Brig. Gen. George O. Squier, the chief signal officer, to prepare the "$640,000,000 Air Bill," which was subsequently passed by Congress. Foulois was definitely not the neophyte that Mitchell made him out to be. Mitchell also characterized Foulois's "staff of non-flyers" as "an incompetent lot" who knew nothing of the European situation, comparing them to a cavalry troop with "200 men who had never seen a horse, 200 horses who had never seen a man, and 25 officers who had never seen either."[25] Nor was Mitchell kind in his characterization of General Pershing, noting the AEF commander "pussyfooted just when we needed the most action."[26] The early months of America's participation in the air-combat phase of the war witnessed conflicting requirements for weaponry and their employment, and a ceaseless clash of personalities, both of which set the stage for a volatile execution of the aviation mission.

Thus, there were major problems on both sides of the Atlantic. Stateside, there were significant management and aircraft production issues, and in France General Pershing took direct steps to solve his aviation mission problems. He knew that the airmen needed to be split off from the Signal Corps if they were ever to be effective. Soon after his arrival in Europe, Pershing created the Air Service, AEF. He noted in his World War I memoirs, "As aviation was in no sense a logical branch of the Signal Corps, the two were separated in the A.E.F. as soon as practical and an air corps was organized and maintained as a distinct force."[27] In August 1917 Pershing demanded and was granted the authority to determine the aircraft types to be procured for the war effort.[28] In September Pershing installed Brig. Gen. William L. Kenly as chief of the Air Service, AEF, and Colonel Mitchell was made the commander for the Zone of Advance. This arrangement lasted until 27 November, when General Foulois and his staff arrived to relieve General Kenly, who returned to the States to replace General Squier as head of the newly created Division of Military Aeronautics in May 1918.

This action came about because of the fiasco associated with aircraft production. From November 1917 through April 1918 a series of headlines detailed General Squier's administration of Army aviation production and procurement as incompetent and rife with conflicts of interest. Squier, who had become chief of the Aviation Section in May 1917, was an enthusiastic proponent of aviation but quickly proved his lack of managerial expertise, especially evidenced by his overly optimistic promises regarding aircraft production. Due to Squier's pronouncements, the War Department released grossly exaggerated accounts concerning U.S. aircraft production. Pershing complained bitterly to the War Department about the grossly excessive aircraft production and employment figures being quoted in American and French newspapers. In reality, as Pershing cabled to General March in February 1918, "there is not a single American-made plane in Europe."[29]

To address this state of affairs, President Woodrow Wilson, by an executive order sanctioned by the Overman Act of May 1918, removed Army aviation from the Signal Corps. Squier was then removed as head of Army aviation, and its mission was divided between two organizations.[30] The director of military aeronautics, Brigadier General Kenly, and his staff were made responsible for training and operations, while the Bureau of Aircraft Production, under John D. Ryan, exercised "full and exclusive jurisdiction and control over production of aeroplanes, engines, and aircraft equipment" for the Army.[31] Unfortunately, the two entities of the new "Air Service, U.S. Army," reported separately to the secretary of war, who at this point was as busy, if not more so, than the president. No single authority was devoted to the administration of a common Air Service policy. Executive coordination between those who made the planes and those who used them was nonexistent. Initially, this shortfall was mitigated by the appointment of Ryan as the "Director of Air Service of the U.S. Army," in addition to his post as the head of the Bureau of Aircraft Production. This arrangement had merit, for Ryan was elevated to second assistant secretary of war, with the responsibility of managing both newly created parts of the Air Service. But this precedent was not codified in law, and when Ryan resigned after the war the post was left vacant.[32]

There was little love lost between Mitchell and the Washington crowd. General Pershing was forced to pick an exceptionally strong administrator to manage his out-of-control airmen. The AEF commander had no choice but to go outside the aviation career field to find his man because the pool of aviators was desperately lacking in command talent and corrupted by uncontrolled egos. Pershing had a war to win and he knew that the aviators could help win it; but Pershing did not, as one historian recently claimed, "give little attention" to his air service.[33] He was far from indifferent to this important combat arm. In fact, he was so concerned with aviation's contribution to the American war effort that he pulled one of his most trusted, dependable, and capable officers out of a critically important command assignment to solve the problem.

When he cajoled Patrick to take the Air Service commander's job, Pershing said, "It needs a strong hand and a man who can see far."[34] At that time, little did Pershing realize just how forceful and steadfast would be the hand and how visionary and innovative would be the eye.

2. Army Engineer from the Hills of West Virginia

Unlike his future assistant, Billy Mitchell, Mason Mathews Patrick was raised far from an affluent cosmopolitan environment. High society did not venture within two hundred miles of Lewisburg, West Virginia. The first time Patrick saw Washington, D.C., was when he was stationed there in 1901. By comparison, Billy Mitchell spent part of his youth amid Washington's federal buildings, attending private school and George Washington University while his father served as a U.S. senator from Wisconsin.

Patrick's family had humble roots that ran back through America's pioneering soil further than the Mitchell family tree. The first Patrick ancestor, Thomas (Kil)Patrick, hailed from Belfast, Ireland, and landed in New England in 1718.[1] Five generations later, Mason Patrick's father was serving as a surgeon in the Confederate States of America army. On 12 December 1863 Maj. Alfred Spicer Patrick rode over one hundred miles to be with his wife, Virginia, when their first child, Mason Mathews, was born the following day. At the end of the Civil War Dr. Patrick had little in the way of worldly goods, but his medical practice assured the family an ever more comfortable life.[2]

The rugged, rural environment of West Virginia prepared Patrick well for a challenging life in the U.S. Army. He attended both public and private schools and, by matter of natural course, his provincial existence contributed to a talent for horsemanship, rough-and-tumble sports, and a life-long love of tobacco, which would ultimately make him the beneficiary of several dozen demerits as a West Point cadet.[3] The rugged West Virginia environment and Virginia's strong military tradition contributed much to the development of a career military mentality.

(The surroundings must have influenced more than just Patrick; less than thirteen miles away from his boyhood home was the homestead of another future general, John L. Hines, destined to be Chief of Staff.)

Patrick was raised in a rather rigorous academic environment; his father prodded him to excel and he did. He was particularly adept at mathematics and gained a local reputation as a young scholar. So exemplary was his academic performance that immediately after his graduation he returned to his high school to teach for two years. Patrick then won an appointment to the U.S. Military Academy at West Point and on 1 September 1882, at the age of eighteen years and eight months, he began his military career.[4]

Patrick's West Point days would not suggest a man later characterized as nononsense and taciturn. Quite the opposite. Cadet Patrick's demerits put him solidly in the Huck Finn category, with repeated infractions for tobacco use, employment of profanity, lateness, and even two infractions for "sliding down the banister." Patrick was cited on twenty-four occasions for being "improperly dressed."[5] His demerits notwithstanding (the total number of which put him in the middle rankings), he excelled academically.

But more important than his class standing were his classmates, among whom was John J. Pershing. Writing in 1927, Patrick described Pershing thus: "General Pershing and I had been classmates at West Point. He was a little older than most of us, more mature, and from the beginning was an outstanding figure in the Corps of Cadets."[6] As third and second classmen, Patrick and Pershing were promoted together to leadership ranks of lance corporal and sergeant. As first classmen (senior year) Pershing and Patrick held the two top posts as, respectively, the first and second captains of the Corps of Cadets.[7] In their four years at West Point, Pershing came to know and trust in Patrick's capabilities.

While at West Point, Patrick eclipsed his classmates in mathematics and engineering, received excellent grades in all other courses, and spoke French rather well. He did not cut a dashing figure, but his intelligence and leadership skills were great assets. As a cadet, he earned positions of authority. He was well read, almost renaissance in nature. Upon graduating second out of seventy-seven from his 1886 class, he was allowed the prerogative of choosing his arm of service. He chose engineering, a prestigious arm, and it determined his career for the next thirty-two years.

Patrick left West Point for Long Island, New York, to serve with the Battalion of Engineers at Willetts Point, where he was under instruction at the Army's postgraduate Engineer School of Application until October 1889. In the summer of 1889, while still at Willetts Point, Patrick was put on detached service to Pennsylvania, rendering aid following the horrific Johnstown flood. Patrick made an impressive showing during his service at Johnstown and was rewarded with the responsibility of designing major flood-control and harbor-construction projects in North and South Carolina.[8] From the Carolinas he went back to West Point where he was an assistant instructor of military engineering, followed by three tours revolving around river and harbor work in Cincinnati, Ohio, Memphis, Tennessee, and St. Louis, Missouri, respectively, until August 1901. On 18 May 1898 Patrick attained the rank of captain during his tour in Memphis. For the next two years he served as an assistant in the Office of Chief of Engineers, Washington, D.C., and then served back-to-back tours at West Point, first as an instructor of military engineering, followed by a tour as commander of the Detachment of Engineers. It was during this tour as commander of the Detachment of Engineers that Patrick had the opportunity to honor one of his relatives, Maj. Gen. Marsena R. Patrick, also a West Point graduate, by overseeing the construction of a fountain in his honor with funds donated by the Patrick family. In future years, after Patrick made flag rank, the maintenance of same was often on his mind when he had occasion to write to the West Point superintendent.[9]

Leaving this West Point assignment as a major in 1906, and having proven his ability to command, Patrick was made chief engineer of the Army of Cuban Pacification. It was during this time that Patrick first met Billy Mitchell. Mitchell's Signal Corps unit was setting poles and stringing telegraph wire on the sides of the roads that Patrick's soldiers were constructing throughout the island. Patrick later noted that his association with Mitchell during this period was amiable and that Mitchell was quite capable.[10] Departing Cuba in 1909, Patrick was put in charge of river, harbor, and fortification works at Norfolk, Virginia, where he pinned on lieutenant colonel's silver leaves in 1910. He then commanded the Detroit, Michigan, Engineer District, which included the maintenance and construction of canals and locks at Sault Ste. Marie. During this tour Patrick was promoted to full colonel on 24 March 1916. Shortly thereafter, Pershing called upon Patrick to command the 1st Regiment of Engineers at San Antonio and Brownsville, Texas, during the Mexican incursion, during which he was appointed as the engineering officer on Pershing's staff. Following the brief diversion on the Mexican border, where his unit saw little action, Patrick was assigned back to Washington, D.C., to command the 1st Engineers as well as being commandant of the Engineering School at Washington Barracks.[11]

Patrick was not in this position very long (nor did he expect to be). Promoted to brigadier general, National Army, on 5 August 1916, the following day he boarded the transport ship USS *Finland* bound for France with the 1st Engineers to support the rapidly expanding war effort.[12] Once in France, Patrick was named, in rapid succession, Chief of Engineering Instruction, AEF, and Chief Engineer for Lines of Communication (later changed to Director of Construction and Forestry). In this latter position Patrick was in charge of all AEF construction work in France, including railways, storage depots, hospitals, housing, docks, airfields, and training camps. The scale of this endeavor becomes clearer when one realizes that these facilities were being built to support over two million soldiers of the AEF, who were rapidly descending on the shores of France.[13]

From August 1917 through early May of the following year, Patrick proved himself of much worth to the AEF commander. This was very evident early on as the Americans were trying to get squared away in France, confronted daily with almost insurmountable logistics and engineering difficulties. Pershing relied on Patrick's judgment and frequently asked his opinion on major issues. Most important, when Patrick did not agree with a particular decision by his former West Point classmate he made his case forcefully but always respectfully. On two particular occasions, Pershing used Patrick in key roles to better organize the AEF. He first called upon Patrick in October 1917 to take over as commanding general of the Services of Supply (SOS) when the original SOS commander, Brig. Gen. R. M. Blatchford, proved unequal to the task.[14] Pershing called on Patrick again when, as chief engineer of the Lines of Communication, he was instrumental in the very successful (and much needed) AEF headquarters reorganization of February 1918. The original AEF headquarters was grossly overstaffed and top-heavy, with fifteen department and service heads in residence, all taking a demanding toll on Pershing's critically limited time. Patrick, Gen. James G. Harbord, and other key commanders discussed the reorganization at length with Pershing.[15] Ultimately, eleven of those offices, with their respective chiefs, were placed under the commanding general of the Lines of Communications, General Harbord. Patrick's contribution to this successful reorganization enhanced his reputation as an excellent administrator.

Late on the afternoon of 10 May 1918, while Patrick was heavily engaged in overseeing all AEF construction requirements, General Pershing's senior aide called to ask if Patrick could be at AEF headquarters the next morning. Gathering up the latest blueprints and status reports, Patrick sped off in his staff car toward Pershing's headquarters at Chaumont.[16] Upon meeting with Pershing, Patrick could not have been more surprised at the AEF commander's request. Patrick described the encounter to his wife, Grace, in a letter he wrote that same evening paraphrasing General Pershing's remarks:

"The fact is I am entirely dissatisfied with the way the aviation service is getting on and I want to put you at the head of it and have you bring order out of what is now chaos, have you manage it and get results. There is bickering, they are running around in circles. There is need for a man to take hold of it and whip it into shape. I want you to do this for me. This is no sudden thing. I have gone over in my mind all the men over here and men who are at home. I know what you have done and of all the men I can get. I am convinced that you are the best one I can find for this job. I know how big it is. I am proposing to give you a huge task. There are questions of policy with the French and English. There should be unity of plan and unity of effort. At present there is nothing of the kind. Someone must bring this about. I think you can do it. You lack only one thing, you are not a flyer, but I do not want a flyer. I can get plenty of them. I want a man on the ground. A man whose feet are firmly planted, one who will know what to do and how to do it."

He told me to think it over, but to remember that he was hard pressed and that he needed help. I was rather overcome—overwhelmed is a better word.[17]

Describing the encounter in his diary entry for 16 May 1918, Patrick wrote, "It staggered me. Nothing to do but try."[18] Writing about the event ten years later Patrick recalled the event much the same as he described it to Grace, but he added: "I . . . knew nothing whatever of aviation. I paid no attention to it even in France . . . being too much absorbed in the work which I was doing and which was along familiar lines."[19]

Patrick was apprehensive concerning his lack of aviation experience and knowledge relative to his new assignment, but Maj. Gen. James W. McAndrew, Pershing's chief of staff, put him at ease. McAndrew let Patrick know that the Air Service individuals who were "running around in circles" were doing so in large part because they all considered themselves to have the right experience and knowledge to run the Air Service, AEF.[20] It must be noted that Pershing's quote characterizing the Air Service as "good men running around in circles" was not aimed at the whole of the Air Service. Pershing was criticizing, in particular, the rather large staff (112 officers and 300 enlisted) that had accompanied Brigadier General Foulois to France in November 1917. With a large contingent of businessmen who had received direct commissions in the Air Service, Foulois had his hands full.[21] Pershing's displeasure with these conditions was still obvious many years later when he was drafting his memoirs of the war. Originally he wrote, "There was perhaps, no branch of the service that gave us more trouble than aviation," but he subsequently changed this sentence to read, "Jealousies existed among them, no one had the confidence of all the others, and it was not easy to select from among the officers of the corps any outstanding executive."[22]

Pershing was not the only one displeased. These newly minted nonflying staff officers displaced many of Mitchell's experienced flying officers in the Zone of Advance, causing Mitchell to characterize the coup thusly: "A more incompetent lot of air warriors had never arrived in the zone of active military operations since the war began."[23] This unilateral usurpation of authority permanently soured the Mitchell/Foulois relationship and dramatically framed the future dealings between fliers and nonfliers in the years to come. Patrick, who "knew nothing whatever of aviation," was headed for a demanding command responsibility.[24]

It was a responsibility that had completely eluded Benny Foulois since his arrival in Europe. While Foulois's association with aviation and the Air Service was a long and distinguished one, Foulois himself was headstrong and opinionated. Like Mitchell he also cared little for all the newly commissioned "upstart" aviation experts and civilian advisors from Washington.[25] But Mitchell (who had been in Europe at the outbreak of the war) and others who arrived soon thereafter thought of Foulois as an interloper. In fact, Mitchell characterized Foulois and his staff as "carpetbaggers."[26]

Be they "upstarts" or "carpetbaggers," the leadership of the Air Service AEF did indeed lead their personnel "around in circles" until, with the appointment of Patrick, they were "made to go straight." It was not difficult to see why Pershing chose Patrick.

McAndrew, shortly after Patrick's appointment as Air Service chief, emphasized that it had taken Pershing some time to find the right individual to do the job that Foulois seemed incapable of doing.[27] But, in all fairness to Foulois, he did have to contend with the obstreperous Billy Mitchell; and in all fairness to Pershing, prior to Patrick's appointment, he desperately wanted the Air Service to succeed with Air Service personnel in the key command positions, and the AEF commander gave the Air Service more than ample opportunity to do so.

The lack of senior officer depth, plus clashing individual egos, put the Air Service on the road to administrative gridlock. But Pershing was prompted by other mitigating factors as well, the most important being that the Air Service had not lived up to its premature claims of aircraft acquisition, as well as its failure to live up to the early trumpeting of the promised effect of American military aviation on the front lines. Patrick's appointment was part and parcel of the total reorganization of the Air Service that President Wilson put into motion in May of 1918 to quiet the accusations of scandal concerning the huge ($640 million) aircraft appropriation program that went awry.[28] In addition, Pershing wanted to name his own Air Service chief instead of being unilaterally dictated to by the General Staff. Last, Pershing was influenced to a greater or lesser degree by various air officers who were caught up in the morass of Foulois's ineffectual staff, and who wanted a change—any change.[29]

Patrick brought change indeed. He made an impressive display of his organizational and management talents when, three days into his new command, he completely reorganized the administrative structure of the Air Service, eliminating redundant and unnecessary management levels and individuals. Within one month of taking charge, Patrick had promulgated twenty-six "memoranda" that covered every aspect of the Air Service organization, to include personnel programs, acquisition and supply projections, training issues, and Allied cooperation arrangements.[30] Some personnel were sacked and some were transferred within the organization. In addition, the initial plan to implement a 386 combat-squadron force, characterized by Patrick as being "absolutely impossible to meet," was revamped to a more manageable 202 squadrons, which he still considered to be too many but at least provided a more realistic benchmark.[31]

Within a week after Patrick took over, Pershing recommended Patrick for his second star, which was approved by President Wilson on 26 June. In a 27 May 1918 letter from Patrick to Pershing, thanking him for his two-star promotion, Patrick's keen management sense is obvious as he characterizes his new organization as having had "too much talk and too little decided action." Patrick expanded on this observation in a lengthy report to Pershing shortly thereafter which was anything but complimentary toward the Air Service.[32]

As Patrick began wrestling with the Air Service, Foulois, at Pershing's direction to Patrick, was placed in charge of all "tactical" matters at the front as Commander, Air Service, First Army. This displaced Colonel Mitchell, who was made commander of the 1st Brigade, Air Service. This decision was not taken lightly. In four separate meetings between Pershing and Patrick between 20 and 23 May 1918, the question of Mitchell's assignment was paramount. Patrick made note of Pershing's decision of 23 May in his diary: "Saw C in C. He decided to give tactical handling of aircraft to Foulois instead of Mitchell."[33]

As an indication of just how bad the relationship was between Foulois and Mitchell, one need only look at the ugly confrontation that resulted during their raucous "change of command." On 4 June Foulois arrived at Mitchell's headquarters at Toul to move into the office spaces occupied by Mitchell. Foulois asked Mitchell that all files, furniture, and equipment remain to facilitate his transition. The resultant shouting match precipitated a 4 June letter from Foulois to Pershing requesting that Mitchell be relieved of command and sent packing back to the States.[34] Pershing gave Major General McAndrews, his chief of staff, the lead in mending what seemed an insurmountable breach. Andrews, by invoking General Pershing's desires, forced the two obdurate officers to come to an accommodation and pledge their respective cooperation.[35] Pershing also took a direct hand: he "talked quite plainly" to Mitchell and gave "him an opportunity to set himself right."[36]

It was highly unlikely that Mitchell would have faced any serious disciplinary consequences as a result of this incident. Pershing respected his enthusiasm and combat leadership capability, as did Gen. Peyton C. March. Perhaps more important, Gen. Hunter Liggett, who at this time was commander of the 1st Army Corps, to which Mitchell was assigned, took a direct interest in his aviation support arm. There is more, though, to the Liggett/Mitchell relationship. In late June Mitchell quickly got on the wrong side of Liggett, who, through a letter written by his chief of staff, Brig. Gen. Malin Craig, severely criticized Mitchell's handling of his recently acquired command of the 1st Brigade, Air Service (1st Corps), for exceeding his authority (ignoring Liggett's staff, and the French staff as well, for coordination purposes).[37] Mitchell's "tactlessness and zeal" brought on an investigation by the Inspector General.[38] In addition, Brig. Gen. Hugh Drum (who replaced Craig as First Army Chief of Staff) was also very critical of Mitchell.[39]

But in short order, with Patrick now at the helm, things improved greatly. In fact, it was after Mitchell's outstanding air-support effort at the Battle of Chateau-Thierry that Foulois requested Patrick's approval for Mitchell to assume his former post as First Army Air Service commander, and that Foulois be named Patrick's assistant. An uncharacteristically magnanimous gesture on Foulois's part, this change took place on 1 August 1918.[40]

Between the time of his appointment as head of the Air Service, AEF, and the end of the war, a brief six months, Patrick's leadership and managerial acumen literally transformed the organization into a model of efficiency. Pershing's faith was not misplaced. But Patrick made decisions during those six months that, even though correct in the context of the time, would nip at his flight boots in his later tenure as chief of the peacetime Air Service. One of those policy decisions concerned who possessed ultimate control over air combat units. In Patrick's view, as stated in his official report after the war:

Air Service Command at the Front: The Air Service was organized upon the principle that at the front it is a combat (not a staff) arm and is to be employed in combination with other similar arms of the Service. The units of the Air Service are organized as integral parts of larger units, divisions, army corps, armies and the G.H.Q. Reserve. They are therefore commanded in the full sense of the word by the commanding generals of these larger units. Responsibility for the performance of the allotted task rests upon the Air Service officer commanding the unit or units involved. The Air Service originates and suggests employment for its units but final decision is vested in the commanding general of the larger units, of which the Air Service forms a part. Commanders of larger organizations exercise direct control over all units, including Air Service units, in their command. There is no separate chain of tactical command in the Air Service.[41]

Regardless of the official chain of command, it was Mitchell who made the vast majority of the operational decisions in the two major AEF campaigns of St. Mihiel and Meuse-Argonne, and this is the way that Patrick, and especially Pershing, wanted it. Pershing was well aware of the dearth of experienced American aviators of Mitchell's caliber; he needed this combat commander. Patrick needed Mitchell, too. It was a wise management decision to keep Mitchell on the front lines. Patrick, with his organizational acumen, devoted himself to the smooth administration of the Air Service as a whole and left operational combat decisions to the fliers, Mitchell in particular.

Patrick visited Mitchell at the front on several occasions and developed a professional—one may even term it cordial—relationship with the charismatic airman that continued after the Armistice. Seeing Mitchell in September 1918, Patrick noted in his diary, "Found Mitchell as pleasant as ever, as full of enthusiasm, but seemed very tired I thought."[42] Patrick respected Mitchell's abilities, so much so that after Mitchell's impressive handling of air assets for the St. Mihiel offensive, Patrick nominated him personally to Pershing for his promotion to brigadier.[43] There is, though, another little-known aspect to the St. Mihiel air victory. The Air Services' stunning performance at St. Mihiel was not due entirely to Mitchell. Shortly after Patrick was appointed chief of the Air Service in May 1918, he instituted two major changes in the areas of maintenance and supply. His first initiative streamlined the logistical system and kept needed supplies flowing to the front and repairable items flowing to the rear. This was done by the establishment of the Coordinating Section at Air Service headquarters that enforced Patrick's supply and maintenance directives and closely tracked personnel and equipment. Patrick also created an Air Service inspection system to ensure strict supply and maintenance discipline in the field. The Air Service success at St. Mihiel was due in large part to a sound logistics system implemented by Patrick that equipped and sustained Mitchell's combat forces.[44] Patrick visited the front-line aviation squadrons for award ceremonies and afterwards sadly reflected in his diary on the unfortunate fate that many of his men met in combat.[45] In fact, squadron commanders wrote many letters of condolence to families back home, as Air Service fatalities were relatively high. World War I Air Service deaths at the front were 152 per 1,000 personnel and behind the lines, 49 per 1,000, for a total of 201 Air Service fatalities per 1,000 personnel. By comparison, AEF infantry deaths were 101 per 1,000 at the front and only 0.441 per 1,000 men behind the lines. AEF artillerymen suffered 13 deaths per 1,000 at the front and 0.572 per 1,000 not on the line.[46]

Air Service, AEF: Mission, Doctrine, and the Future

Patrick was very much a team player, being quite egalitarian in his views of who contributed what to the mission as a whole. But he did feel strongly that his observation personnel had been treated unfairly, both on the Western front and on the home front. In his end-of-tour report, Patrick commented on the value of reconnaissance: "the work of the observer and the observation pilot is the most important and far-reaching which an air service operating with an army is called upon to perform." Almost in the same breath, Patrick took a swipe at how the colorful and romantic depiction of the pursuit mission "proved a serious handicap to the development of other branches of the Air Service . . . with the inevitable result that observation pilots and observers lost caste among their fellows." While he decried the unfair characterization of the observation branch, Patrick still complimented the capabilities of his pursuit pilots who had "reached a stage at which [they] ranked in efficiency with the pursuit aviation of the Allied Armies . . . with pilots second to none."[47]

It is a telling indication of Patrick's quick mind that he so rapidly assimilated and embraced many of the basic notions of early air power. While noting the key importance of reconnaissance work in his final report to Pershing, often overlooked is what Patrick had to say about the bombing mission. Patrick knew that early bombing efforts by the Allies, and then by AEF Air Service bombardment groups, led to heavy losses of aircraft. Patrick took three key initiatives to better protect these planes: an emphasis on aerial gunnery; tight formation flying; and cooperation with pursuit aircraft to fly cover and protect the bombers. In retrospect, the answer to World War II bomber force protection lay in the experiences of the World War I Western front and, more important, in the experiences of those same pilots who went on to key command positions in World War II. In the words of General Patrick:

The utmost stress was laid on gunnery . . . formation flying was insisted upon . . . a tight formation meant safety . . . objectives were attacked by a whole group instead of a single squadron . . . better cooperation was secured with pursuit. . . . This reduced our own losses and increased those of the enemy.[48]

This was the creation and codification of doctrine. Air Service combat doctrine simply did not exist prior to this conflict. During the war Mitchell and company were basically making up the doctrine to meet the war's demands—almost all based on British and French practice. To be more precise, Mitchell borrowed operational doctrine from the British, Italians, and French. He contributed nothing new. Mitchell simply transferred these Allied ideas to the operational requirements of the Air Service. There was an obvious benefit: it facilitated joint operations.

In hindsight it is quite amazing that prior to World War I there had been no attempt to consolidate doctrine as, in the words of one historian, "almost nobody in the Air Service seemed to be aware of the need for an objective, systematic, and sustained study of concepts of employment and tactical methods of operations."[49] Captain Foulois had written up some proposals for the use of aircraft in support of combat operations as did Capt. H. H. "Hap" Arnold, but they were a far cry from any systematic study. Prior to the war, the Signal Corps, with its Aviation Branch stepchild under its control, cared little for the promulgation of what could become an expensive failure, or an expensive success. Regardless, when war did come, Pershing removed all aviation units in France from Signal Corps control. This action, combined with the mismanagement of the aircraft production program, resulted in the complete removal of aviation from the Signal Corps. In essence, this was the beginning of aviation's emancipation, which Patrick, in the 1920s, further enhanced.

But a relative degree of organizational freedom for American military aviation did not precipitate a concomitant initiative with regard to doctrine that would validate any increase in independent action. One could write off this neglect due to the exigencies and demands of the war, but that does not explain why the immediate postwar Air Service sorely neglected the development of much-needed doctrine to justify their awesome requirements. When one looks at the tremendous growth of the air arm, from several dozen men and a pair of aircraft to over 150,000 men and 15,000 aircraft within the space of ten years, it gives one pause when one considers the almost total disregard for a complementary doctrine of employment. What makes this issue so important is the central role that it played in the resource and independence controversy involving the Air Service in the 1920s when Patrick was at the helm. An eminent historian who has studied the question of early air doctrine later wrote:

The United States entered the war without a clearly defined doctrine of aerial warfare . . . [and] the Air Service manifestly did not learn from wartime experience the critical importance of systematic formulation of doctrine as a step essential to successful development of air weapons. In consequence, the growth of the air arm in peacetime suffered a significant handicap.[50]

This characterization is decidedly true of the immediate postwar Air Service under the tenure of Maj. Gen. Charles T. Menoher, but not an entirely true reflection of the service under the leadership of Patrick, who took over in October 1921. Personalities and politics were more to blame for the lack of coherent air doctrine in the immediate postwar years than anything else. The World War I air-war tactical "lessons learned" were learned well; they were not so much forgotten as they were conveniently ignored by the powers who controlled the purse strings, both civilian and military. Below we shall see how, in time, Patrick became assertively proactive in his approach to utilizing the air-power doctrinal lessons of World War I to justify his fight for funding and a greater degree of autonomy for the Air Service.

3. Patrick Wins His War within a War

Patrick faced many problems during his six-month tenure as head of the Air Service, AEF. How he approached and resolved these issues is but a brief prelude to the even greater difficulties he would face when Pershing again asked him to head the problem-plagued service in 1921. What is notable about Patrick's World War I service is not his administrative acumen as an engineer or as Air Service chief but his adaptability under the pressures of war. Thirty years an engineering officer turned overnight into head of the AEF's aviation effort was one thing, but to take on such a radically unfamiliar mission bedeviled by personality, training, Allied coordination, and supply issues demonstrates the mark of true leadership and command genius. Whatever the issue, Patrick did not act arbitrarily or personally; he acted in the best interests of the AEF war effort.

For Patrick it was most fortunate that the best interests of the AEF at this point did not concern the battle for funding or for autonomy: the staggering sum appropriated for aviation during the war was well over $1.5 billion, and independence would not become an issue until after the war.[1] Patrick was tasked by Pershing to put the fliers on a true and stable heading by quelling personality conflicts and giving organizational direction to the service. This he did. It was unfortunate that his abilities were not put to use putting the stateside aircraft production problem in order, for it would have direct bearing on the condition of the Air Service in the 1920s.

Even with huge appropriated sums, the American aircraft industry could not solve the myriad problems that existed. The automobile industry could not take up the huge slack either. Eventually the first American-built DH-4 was shipped from

Hoboken, New Jersey, in March of 1918. Ultimately, only 196 of these aircraft made it to combat squadrons by the end of the war. In September 1918, when Billy Mitchell managed so brilliantly the 1,481 aircraft in support of the St. Mihiel salient campaign, only 609 of the planes were from U.S. squadrons, and but a fraction were American-made.[2]

It was only as the war was drawing to a close that American manufacturers began to produce respectable quantities of planes: over one thousand DH-4s were manufactured in October and November 1918.[3] In retrospect, given the state of the American aircraft industry prior to World War I, it is an amazing accomplishment. Prior to America's entry in the war, there was little association on any level between Americans and the belligerents due to President Woodrow Wilson's desire for strict neutrality. In addition, when it came to technical expertise, American aircraft manufacturers were far behind their European counterparts and lacked a manufacturing facility that could produce in quantity.[4] Patrick was well aware of the inadequacies of America's aircraft manufacturing effort. He was also cognizant of Pershing's critical view of the Aircraft Production Board (September 1916), and its replacement, the Aircraft Board (October 1917), and, finally, the Bureau of Aircraft Production.[5] Pershing's wrath was also targeted at the Signal Corps' inept administration of the aircraft acquisition program. Pershing and Patrick were not so much critical of the lagging airplane manufacturing effort as they were about the extravagant stateside claims being made by civilian and Signal Corps personnel associated with the aircraft production program. As Patrick described it:

There had been, especially in the United States, most extravagant statements made of what we would do in the air, how we would fill the air with planes, and how we would overwhelm the German airmen. It was popularly supposed at home that almost from the very beginning these fleets of American planes would be dealing death and destruction to the Germans, and yet more than a year after we entered the war not a single American-built airplane had appeared in France.[6]

Overall, the Air Service, AEF, reaped the bitter fruit of congressional, General Staff, and Signal Corps neglect. In addition, it suffered from the severe strains of many disruptive fits and starts prone to new organizations born under stress. The air arm also had difficulty living down its abysmal showing in the Mexican Punitive Expedition. Still fresh in many minds was how little the fragile airframes contributed to the mission. Foulois had commanded that air effort. Pershing, who commanded the expedition, remembered.

Patrick, on the other hand, had always retained the confidence of the AEF commander and was given assignments that validated that trust.[7] Dealing with the

British was especially taxing. While Patrick had developed a good professional working relationship with Maj. Gen. Hugh Trenchard, commander of the Royal Flying Corps in France and later commander of the British Independent Air Force, his dealings with London officialdom were less than cordial.[8] Patrick made several trips to London to assuage the British government's wholly unfounded concern that the Air Service was falling under the advisory spell of the French. He also negotiated for the "release" of thousands of American airmen who were in training there, being used by the British as surrogate manpower for their undermanned aerodromes. At this time, over 15,000 U.S. Air Service personnel were in England. Patrick was determined to obtain the release of at least a moderate percentage of these men. During one of his trips to London in June 1918, Patrick met with Lord Weir, the Air Minister, for just that purpose. Weir was anything but cooperative. He began the meeting by detailing his exasperation with the truculent and ever-changing Washington bureaucracy, then rapidly switched to the availability of Liberty engines, and then wanted to know the status of the American relationship with the French. Patrick waited patiently, then calmly drove home his point regarding the transfer of England-based Air Service personnel to France. Eventually he convinced Weir to release 3,500 men.[9]

Under Patrick's adroit administrative leadership and Mitchell's inspiring and skillful combat leadership, the Air Service, AEF, evolved into a capable, albeit relatively small fighting force. Ultimately there were forty-five American squadrons, with 740 airplanes manned by 767 pilots, and 481 observers.[10] While only twelve of the squadrons were equipped with American-built aircraft, literally thousands of planes in the production pipeline were headed toward the front.[11] American pilots and observer/gunners brought down 776 enemy planes, while losing 290 of their own aircraft for a better than 2.5 to 1 kill ratio.[12] As could be expected, the American bombing effort was much less impressive: 138 tons of bombs on 150 targets, of which very few were at the deepest penetration of 160 miles behind enemy lines. Far less impressive was the record of the Night Bombardment Division, though certainly not for lack of effort. It was simply doomed for lack of bombers. On 11 November 1918 only eleven sets of Handley-Page bomber parts had been received at the designated British assembly plant.[13] This lack of coordination and failure on the part of the American aviation industry did not go unnoticed by Patrick.[14]

General Patrick and the World War I Strategic Bombing Issue

During his tenure as chief of the Air Service/Corps, Patrick consistently harped on the aviation industry's ability to transition from a peacetime to wartime footing. "You cannot build it overnight" was his main theme. Had the requisite amount of industrial aviation infrastructure and capacity been available, aircraft would have been at hand to execute missions and provide a body of strategic aviation experience to guide the Air Service following the war. At the close of the war prescient Air Service theorists (and very few there were) discerned the promise of strategic military air power, but these theorists lacked a base of experience. Nothing could be quantified.[15] This lack of operational experience was crucial to the postwar controversy regarding the efficacy of air power, and Mason Patrick would be swept up in the debate. In fact, Patrick has been unfairly criticized for his seemingly lukewarm, if not hostile attitude concerning the strategic air mission and, thus, by faulty logic, he has been characterized as being opposed to Air Service independence. Put into proper context, nothing could be further from the truth.[16]

Patrick, in the midst of war, was tasked by Pershing with bringing order to the chaotic Air Service.[17] He pushed basic priorities that were far more important than the minority-touted strategic bombing mission: first, an all-important revamping of Air Service organization, and then solving the immense problems associated with training, reconnaissance, and fighter aircraft acquisition. Patrick then moved on to the construction of adequate Air Service facilities and the negotiation of bilateral agreements with the British and French air services. His managerial finesse permitted the Air Service, AEF, to fly and fight to a degree that would have been impossible under the early Air Service regimes.

On the surface, Pershing left no doubt concerning his feelings regarding AEF independence. In January 1918 Major General Trenchard suggested that the Air Service, AEF "come in with us [the RAF] as regards [long-range] bombing from the Nancy region." Trenchard's deputy, Brig. Gen. Gerald Blaine, who presided at a subsequent meeting with the AEF commander, described Pershing's reaction thus: "I could see clearly, and in fact he said so, that he is not at all desirous of putting American personnel under us, partly because he wants to keep independent, and partly owing to the trouble it might cause with the French." Blaine's memo on this meeting ended with the observation that the Americans would "run a separate bombing show."[18]

What is most significant about this meeting was Pershing's complete ignorance of Trenchard's joint long-range bombing proposal. It caught Pershing totally unaware. Foulois had simply not notified Pershing of Trenchard's two-week-old memorandum suggesting the arrangement. A 14 January 1918 letter from Trenchard to Pershing, and Pershing's reply of 6 February 1918, set things right: Pershing promptly approved of the plan.[19] Sloppy staff work such as this certainly did not enhance Foulois's reputation on the General Staff, and it was one more reason Pershing finally selected Patrick for the Air Service job. Called to task for this oversight, Foulois demurred rather weakly that he thought the memo "was for his own information only." Pershing had not been served well by the Air Service staff up to this point. Upon assumption of command, it took Patrick less than a month in the job as Air Service chief to get his own staff in line and bring the British Air Staff to task to explain the annoyingly complex arrangements between the two services "so we can know precisely where we stand."[20]

Just where then did Pershing and Patrick stand on the issue of American Air Service involvement in long-range or strategic bombing? Pershing heartily endorsed Trenchard's combined bombing operation plan, stating: "you may be sure that I shall do everything in my power to make this co-operation as effective as possible."[21] Patrick has been unfairly maligned as one who was adamantly opposed to such a mission. If Patrick felt as such, he certainly would not have supported American efforts in this arena, even to the extent of supporting long-range and interdiction night bombing efforts. The limiting factor was simply the lack of available bombers. What Patrick saw as far more important was supporting the troops on the ground with the Air Service assets that were readily available to do the job. He saw the abysmal progress made on the American manufacturing front to produce the wholly unrealistic 358-squadron air force that was envisioned when America first entered the war. Patrick was incensed by the graft and corruption associated with that production effort.[22] He used the assets he had on hand and those he could reasonably expect to receive to accomplish the most pressing mission at hand. When the short- and long-term airplane manufacturing picture began to brighten during October 1918, Patrick's staff did not hesitate to task Brig. Gen. Dennis E. Nolan, the AEF G-2, with gathering information on German industrial targets "for the purpose of arranging a program of airplane bombing." Nolan requested detailed target information from the British and Trenchard wasted no time honoring this request.[23]

In relation to the issue of strategic bombing, it must also be noted that in early May 1918 the Inter-Allied Aviation Committee was formed (an idea first ventured by Foulois in January 1918).[24] This joint committee was chaired by Maj. Gen. Frederick H. Sykes, the British Air Ministry's ranking military member. The main topic under discussion was a British-proposed Inter-Allied Bombing Force. Foulois, the American representative, with General Patrick's approval consistently voted in favor of the plan. (The French consistently voted against.) This issue made its way up to the Supreme War Council (SWC), where Gen. Tasker H. Bliss was the American military representative. Bliss voted in favor of the Inter-Allied Bombing Force, signing the memorandum that stated the organization would:

be in a position to carry out powerful and intensive bombardment, from the air on enemy territory, both with the object of destroying military objectives and, in case of need, to execute reprisals in German towns [and] to begin at once the elaboration of a methodical plan for the bombardment of towns and industrial centres belonging to the enemy.[25]

Bliss, in his previous capacity as Army chief of staff and then as the U.S. representative to the SWC, in almost every regard had deferred to Pershing's wishes. Now, having approved of the joint strategic bombing plan, Bliss gave America's official sanction to it as well. Patrick was also well aware of and supported the Inter-Allied Bombing Force and, far from taking "umbrage" at the idea of strategic bombing, he appreciated its capabilities but was realistic about its limitations.[26] Throughout Patrick's career, he held a keen appreciation for what was practical as well as keeping an open mind to the future promise of new technology.

The roots of the strategic bombing controversy of the interwar period begin here, but only insofar as the concept was proposed and operationally tested, albeit on a very limited basis. At this point, Mitchell was not an advocate of strategic bombing. His final report at the end of the war is significant for the fact that air superiority and observation are the most critical accomplishments. It was not until Mitchell's book, *Winged Defense,* was published in 1925 that he emphasized the importance of strategic bombing.[27] Mitchell was, however, an exceptionally capable tactical air combat commander.[28]

While German airmen rightfully receive the credit for pioneering long-range bombing techniques, the British deserve credit for establishing an independent air force driven by the need to improve on the use of air assets to take advantage of the efficencies of centralized command. This was one of the key issues justifying the formation of the independent Royal Air Force via the Smuts Commission of 1917. The other two reasons were to ensure the air defense of Britain and to bomb Germany.[29] Thus, during World War I the British War Cabinet assented to the formation of the Independent Air Force (IAF) for strategic bombing under the persistent prodding of Lord Weir in answer to the panic induced by the Gotha bomber scourge. Lord Weir ordered General Trenchard to establish such a force. Initially Trenchard demurred, even calling it a "great waste" and not wanting to split his force; he argued forcefully for unity of command but to no avail.[30] The political will and priority won out, and the IAF came into being on 5 June 1918 with Trenchard at its head. While Sykes was the main proponent behind the strategic bombing operation, the issue was not new to Trenchard.[31] He had been planning for it since early 1915, but he wanted it done as part and parcel of an independent Royal Air Force.[32]

When the joint strategic bombing issue was ultimately put to Pershing by Trenchard, the AEF commander supported it without reservation. Foulois emphatically endorsed the issue, and Patrick did as well.[33] Patrick not only supported the strategic bombing effort but also had a keen appreciation for centralized command of tactical air assets. When he heard a rumor that the British were entertaining the idea of dedicating "an observation flight or a reconnaissance flight to each division" he directed that this matter be investigated thoroughly. No less a figure than

Maj. Gen. J. M. Salmond, who took over for Trenchard as Commander, HQ, RAF, in the field, replied to Patrick that "we have no intention of assigning Squadrons and Flights to Divisions" and went into detailed justification, emphasizing "the disadvantages are that the system is wasteful in personnel and machines."[34] This lesson was not lost on Patrick, who subsequently strongly supported the concept of centralized command as employed by Mitchell at the upcoming St. Mihiel engagement. Patrick also praised Trenchard's approach to airpower employment and thanked Trenchard for his suggestions and guidance when Patrick assumed command of the Air Service.[35] Mitchell was not the only Air Service officer schooled at the altar of British airpower theory.

But even though Patrick appreciated and supported the concept of centralized control of air power, he stated in his final report that the ultimate control of combat units remained with the Army element to which they were attached: "The Air Service originates and suggests employment for its units but final decision is vested in the commanding general of the larger units, of which the Air Service forms a part.... There is no separate chain of tactical command in the Air Service."[36]

Although these two philosophies may seem at odds, in operational reality Air Service combat units, and in particular Billy Mitchell, exercised considerable independence in the choice of objectives.[37] While Patrick may have endorsed the Army party line in his final report, in actuality he endorsed this rather flexible approach to the use of air power when he fully supported Mitchell's efforts in preparation for, and execution of, the St. Mihiel offensive.

Mason Patrick, Billy Mitchell, and the AEF Staff

When the earliest AEF Air Service and AEF general headquarters staffs were being formed, (then) Col. Billy Mitchell did little to inspire the confidence and trust of newly arrived General "Black Jack" Pershing or the trust of Pershing's key staff members.[38] Based on his staff's recommendations, Pershing allowed Mitchell to be placed "in charge of tactical aviation when necessary," a rather lukewarm endorsement.[39] While Pershing was successful in preventing the dismemberment and piecemeal utilization of American doughboys, regardless of Allied blandishments and bluster, Mitchell found much to be admired in the way the French and British aviators fought this war. Indeed, Mitchell was an Anglophile and was much taken by the commanding presence of Maj. Gen. Hugh Trenchard, who, as commander of the Royal Flying Corps in France, convinced Mitchell of the absolute necessity of the offensive and strategic nature of air power.[40] The especially pro-British attitude that Mitchell displayed was all the more reason that Pershing and Patrick did not view Mitchell as a complete team player, a quality Pershing valued highly. One cannot fault him for emphasizing such a trait, given the almost overwhelming challenges facing the AEF commander: "loyalty . . . was a cardinal virtue in Pershing's hierarchy."[41]

Indeed, it seems Mitchell went out of his way to offend many individuals in key positions who could have been of immense help to him. Among Mitchell's detractors were Pershing's one-time chief of staff, Maj. Gen. Hugh A. Drum, and Brig. Gen. Dennis E. Nolan, the AEF's G-2.[42] Within the Air Service itself there was no lack of Mitchell detractors. First and foremost, of course, was Brig. Gen. Benny Foulois. Col. Frank P. Lahm, a member of Foulois's staff, was a veteran aviator who flew with the Wright brothers and was eventually chief of Air Service, Second Army. He was not enamored with Mitchell either. On 2 June 1918, when Foulois bumped Mitchell out of his position as commander of Air Service, 1st Army, Lahm characterized the Foulois/Mitchell switch as "a good thing. . . . Someone [else] should have been appointed to that position long ago."[43] Even Col. Edgar S. Gorrell, an early friend of Mitchell (who wrote the famed World War I Air Service histories and a seminal study of the American bombardment situation in France), gained Mitchell's enmity when, as a member of Pershing's staff, Gorrell pulled no-notice inspections of Mitchell's squadrons.[44] On the other hand, Mitchell had the support of Hunter Liggett, former commander of the I U.S. Corps, and U.S. First Army, whom Mitchell had ably served as I Corps Air Service Chief.

At the close of the war, Patrick, in writing to Pershing, characterized Mitchell as one who

thinks rapidly and acts quickly, sometimes a little too hastily. He is opinionated but I have usually found him properly subordinate and ready to obey orders. While he has worked well with the men and material which it was possible to furnish, his own ideas of what were necessary to accomplish his tasks I have found sometimes exaggerated. In other words, he has asked for more in the way of personnel and transportation than I believe to have been absolutely necessary for the performance of his duties. He has some tendency to act on his own initiative; it is not meant that this is a fault, as it is frequently a virtue, but there have been a few times when it has been uncertain just where he was or what he was doing. He is at all times enthusiastic and full of energy.[45]

The words that Patrick used to describe Mitchell at the close of the war can very well be applied to how aviation advocates were viewed as a whole: opinionated, exaggerated, and uncertain of just where it was all going. It was personalities, though, which drove those perceptions to a great extent. But what about the results of the war in the air over France? It has been argued that the mission of aerial reconnaissance (the major mission of the air war) for artillery spotting actually prolonged the war by "inhibiting movement . . . and contributing to the defensive side of warfare."[46] What, then, could decisively overcome the stalemate of trench warfare? In the minds of several outspoken aviation advocates, strategic bombing seemed to hold the answer. But as with all past assurances regarding the capabilities of aviation, there were those who fervently believed and promised, and there were those who saw only the failures to measure up to the many promises.

Billy Mitchell came out of the war a fervent believer in the capabilities of air power and he promised much. Patrick came out of the war much as he had gone in, a talented administrator and leader striving to overcome the inertia and personalities within a conservative bureaucracy to accomplish the mission.

The World War I relationship between Patrick and Mitchell was important and symbiotic. In the management of the AEF Air Service between May and November 1918 Patrick and Mitchell ran the air show. Mitchell orchestrated the combat portion and Patrick provided the overall administrative framework of support services: logistics, personnel, supply, training, quarters, and so on. It was a very successful relationship: both were expert in their respective fields, and both contributed their exceptional individual talents to help win a war.

Patrick, after reassuming his role as chief of the Air Service in October 1921, also resumed his professional relationship with Trenchard and corresponded with him intermittently as the need arose.[47] His first time as chief of the Air Service, Patrick had inherited an organization of poor operational repute. But there was a bright spot: the exploits of the Western front fliers were lionized in the press. At the same time, many newspapers savaged the inability of manufacturers to produce the large numbers of promised aircraft as outlined by the Ribot cable and the Bolling Commission. The Bolling Commission of June–July 1917 was tasked with deciding on the correct mix and type of aircraft that America would contribute to the Allied war effort. Initially enamored with the promise and future possibilities of bombers, the commission ultimately recommended that the DH-4 be emphasized for reconnaissance purposes. Practical considerations drove this decision: American industry could not accommodate the aircraft types and their ratios desired by the General Staff and by Pershing: both emphasized pursuit aircraft over observation and bombardment in ratios of 5 : 3 : 1 and 6 : 4 : 3, respectively.[48] Pursuit (fighter) planes designed and built in America would have been woefully obsolete by the time they reached the front.[49] The air arm was to be oriented toward ground support; and the airplane that carried the load in this endeavor was the DH-4.[50] Over 1,200 of the 3,400 DH-4s built in the United States during the war were shipped to France.[51]

The DH-4s on stateside training duty, plus several thousand in the pipeline at the conclusion of the war, plus the thousands of crated spare Liberty engines, kept the Air Service and Air Corps in airframes well into the mid- to late 1920s. Being stuck with a less-than-optimal plane in bountiful quantities proved to be a bane to Patrick's attempts to modernize the service during his 1921–27 tenure. The future Air Service/Air Corps was saddled with this problem because of the exigencies of war and practical considerations of what American industry could accomplish faced with such demands. Patrick, as head of the Air Service in the early 1920s, kept these lessons in mind and took the lead to lobby Congress actively to implement policies that would help to optimize all facets of America's aviation manufacturing capabilities, especially in peacetime.

Prior to Patrick's appointment as chief, Air Service, AEF, the service was indeed rife with conflict. The personal antagonisms, though still simmering under the surface, did not again break out into the open for the duration of the war, although in one instance several officers who were among those who were made to "run straight" when Patrick took command complained to John D. Ryan, the director of the Air Service, during a cursory inspection of Air Service activities in France. Ryan told Pershing that according to some Air Service officers Patrick's management style had several shortcomings. Pershing quashed such a notion on the spot.[52]

Pershing appreciated Mitchell's combat leadership but was frustrated by his management flaws during the war. Pershing had no such qualms about Patrick's performance. Patrick recreated and ran an efficient and effective military Air Service organization that transported appropriate personnel who were trained, equipped, and molded into effective combat squadrons for front-line duty. From being totally unprepared for war in April 1917 to being heavily engaged in battle one year later, Pershing built a successful command and staff organization and two combat capable armies under U.S. command in remarkable time.[53] Mason Patrick deserves credit for relieving Pershing of the administrative nightmare that was the Air Service prior to June 1918, and Patrick's firm management permitted the Air Service to contribute materially to the ultimate Allied victory. For this accomplishment, at war's end he personally received the Distinguished Service Medal from General Pershing.

4. Patrick and the Postwar AEF Air Service, 1918–1919

Pershing appealed to Patrick to remain in Europe after the war ended, and he did so until July 1919. If Pershing needed Patrick's administrative skills during the war, he perhaps needed them even more so during the transition to peace. Pershing wanted Patrick to arrange the return of Air Service personnel to the States, attend to the disposition of Air Service contracts and equipment, coordinate the assignments of Air Service units to postwar roles and missions on the continent, and serve on several treaty-related committees.

The planning for some of this was actually begun about a month prior to the Armistice. On 13 October 1918 General Bliss called Patrick in and explained that a "Committee of Peace Mission" (subordinate to the Supreme War Council) had been formed to consider postwar aviation matters. Bliss tasked Patrick with developing a statement describing the "Air Force Germany should be allowed to maintain based on ground army permitted." Patrick submitted the plan to Bliss on 17 October, noting with a rather cavalier remark in his diary that "I cared not whether he liked it." Bliss must have found favor with it, as did Pershing, for it was presented at a 20 October meeting presided over by the Allied Commanders-in-Chief: Marshals Foch, Petain, and Haig, and Generals Duval, Weigand, Pershing, Diaz, and Grove, with an assortment of admirals and Japanese representatives as well. The meeting was called to discuss the terms to be imposed upon Germany in respect to air forces. According to Patrick, Foch dominated the meeting. Although there were no immediate resolutions, Pershing and Bliss ultimately appointed Patrick as the U.S. representative to several committees dealing with military and commercial aviation issues relating to the peace treaty, primary of which was the Aeronautics Commission of the Supreme War Council.[1]

This period of Patrick's career one can best characterize as looking forward to retirement. His belief that he had experienced the zenith of his career is manifest in the almost whimsical way he approached his duties, and in the way he characterized the many meetings and social events where he rubbed very famous elbows. His diary during this period describes the more prosaic aspects of his routine, from the daily weather to French architecture and cuisine. But to some habits Patrick remained addicted. This was especially true when describing the various individuals with whom he dealt. He pulled no punches.

One of the more entertaining and enlightening stories regarding Patrick's character deals with Adm. Silas Knapp. Knapp and Patrick were appointed to a commission to draft a convention to govern international flying, as well as the consideration of aviation questions forwarded by peace conference members. Patrick's diary description of Knapp gives a clue to Patrick's feelings about not suffering fools gladly and about the U.S. Navy in general:

Admiral Knapp is a funny little man. Knows nothing of the matters coming before the conference. A stickler for rank, sprang it on me. Wanted to run everything. Came up to my office and began to talk about the order of our sitting around the table. Whether I should be on his right or left and spent half an hour mulling over the matter. I told him I did not give a whoop one way or the other . . . with due respect to his exalted position. These things are annoying. Just like all Navy men I have come across.[2]

In reality, personalities aside, the issues that Patrick dealt with during this period would serve him well when he again found himself as head of the Air Service. Patrick drafted the peace treaty terms governing the surrender of German aviation materials and "prescribing the air activities which would be permitted in Germany after the signing of the Peace Treaty."[3] Arguably, more important than the military aspect of Germany's aviation status after the war was Patrick's involvement in the peace treaty's treatment of commercial aviation, where he was schooled in every aspect of what would become a large part of his job as chief of the Air Service and Air Corps between 1921 and 1926. Along with other military and civilian aviation experts, Patrick helped draft all aspects of the commercial aviation part of the treaty. He was well aware of the future "importance which aircraft would play in commercial transportation."[4] Many of the commercial aviation issues (navigation aids, air routes, landing rights) which Patrick dealt with at this time prepared him for his dealings with manufacturers, bureaucrats, and politicians when he led the Air Service again.

Another activity at this time that helped prepare Patrick for future Air Service responsibilities was the Air Service contribution to the Dickman Board. On 19 April 1919 Pershing convened a board of superior officers in Paris, under Maj. Gen. Joseph Dickman, to review the respective findings forwarded by boards of senior officers tasked to review the performance of each branch of the AEF, including the Air Service. Pershing charged its members to "consider the lessons learned during the war insofar as they affected tactics and organization."[5] A 5 May memorandum subsequently directed Patrick to provide "a report upon the present organization and operation of the Air Service and recommend any desirable changes therein." Patrick appointed Foulois to head the Air Service Board and the subsequent report was forwarded to Pershing on 19 May. There were no surprises in the relatively conservative report save two. Included was a minority position (signed by Foulois) suggesting that the chief of the Air Service be a member of the General Staff. Patrick did not concur, stating that regardless, "the Chief of the Air Service will still be held responsible for all matters relating to the efficient operation and maintenance of that service."[6] Additionally, the report emphasized that the prime mission of the Air Service was the collection and transmission of information, and that bombing of distant targets was a "luxury." Patrick's dissent on this second issue is evidence of an open mind regarding the Air Service's future capabilities: "When it is possible to place . . . a bombing force in the field, its size should be limited only by the nation's ability to provide it and by the number and importance of the enemy activities which are to be attacked."[7]

The final Dickman Board results, and the Air Service Board report that contributed to it, plus Patrick's review, were pretty much in harmony as far as the future status of the Air Service was concerned. All supported the status quo: Air Service missions and units were integral parts of the Corps and Army structure, with authority vested in the ground-force commander.

The timing of the Dickman Board, though, was curiously coincident with the impending arrival of one bureaucrat who made Patrick especially anxious to get home and "fill a less exalted rank." In May 1919 Benedict Crowell and the commission that bore his name arrived in France. This mission, or more formally, the American Aviation Mission, was sent by Secretary of War Newton D. Baker on a tour of Italy, France, and England to survey the status of current and projected aeronautical development. Crowell headed the mission with representatives from the Army, Navy, and the aviation industry. What is curious about this mission vis-à-vis its relationship with General Patrick is simply that there was no meaningful relationship or dialogue between the two men. Crowell had only one official meeting with Patrick. In addition, it was also a matter of personality. Not only did Crowell and Patrick not get along, Pershing himself thought little of both Crowell and another prominent member of the mission, industrialist and aeronautics booster Howard E. Coffin.[8] It is easy to surmise why Pershing was not enthusiastic about Coffin's presence. Just prior to America's entry into the war, Coffin had testified strongly in favor of the consolidation of all facets of American aviation into a Department of Aeronautics, removing aviation entirely from the control of the Army. This did not sit well with Pershing, the ultimate team player. In addition, as an aviation "booster," Coffin had found himself embroiled in the midst of the recently investigated aircraft production scandals. This, no doubt, colored Pershing's and Patrick's perceptions. With regard to the timing of the Dickman Board—which was accomplished in record time—and the arrival of the Crowell and Coffin crowd, it can only be surmised that Pershing, upon learning of the impending American Aviation Mission (due to arrive in three weeks and whose members were, for the most part, vocal aviation advocates), sought to dampen their ultimate effect.[9] Whether Secretary of War Baker prompted Pershing to initiate the Dickman review is open to question, as there is no documented proof. But what is documented is Baker's very strict instruction to the independent-minded Crowell "to limit himself to fact-finding and submit no conclusions as to air policy."

Patrick's feelings about Crowell and company left no room for debate: "Assistant Secty War Crowell and his gang including Coffin around. I am sure they do not know what they are going to do or how to do it. Just another joy riding bunch of Americans jaunting about . . . at government expense."[10] At a 5 June meeting with Patrick, after stating he "knows nothing of the Air Service," Crowell asked Patrick for a plan for the postwar Air Service, and Patrick complied on the twenty-third. On 21 June, though, Patrick briefed Pershing on the memo he would soon pass to Crowell. Although Patrick asked Pershing about his views on the matter, Pershing replied that he "did not want to take any decided stand now about the Air Service organization in the States." Patrick, commenting on this meeting, noted that Pershing, while deferring comment on his feelings at this time, freely encouraged Patrick "to give to Assistant Secretary Crowell my personal views [on the Air Service reorganization plan]."[11] As might be expected, Patrick's recommendations to Secretary Crowell for the postwar reorganization of the Air Service did not include a recommendation for immediate independence. Patrick was much too politically and institutionally astute to suggest something so drastic, and, quite simply, he saw no need for independence.

After giving Crowell his planning memorandum, Patrick's bitterness at having experienced the assistant secretary's condescension was obvious in the way he characterized the American Aviation Mission in general, and Crowell in particular: "I have seen little of the said Assistant Secretary. . . . I fancy Coffin has told him he need pay no attention to me. I care little."[12] Regardless of his disclaimer to the contrary, this obvious slight doubtless pained Patrick, and he ended his last days in France very mindful of the fact that he was about to "drop at once from [his] high estate."

Homeward Bound to Retire

The infighting and politics within the Army that Patrick witnessed at this time would also prepare him for his later travails. Jockeying for postwar position was a keen art form. Evening dinner conversations with various colleagues invariably turned to the Army's postwar organization and who would be picked for which command. There were those individuals obliquely asking about Patrick's plans for the future, as Patrick so aptly penned in his diary: "sounding me out about my ambition to be chief." There were those who pulled Patrick aside and in hushed tones under the swirl of cigar smoke noted that Harbord and Pershing were recommending them as head of the Air Service.[13] Patrick, though, entertained no thought whatsoever that he would ever again be associated with America's military air arm. He was unequivocal in a 27 June 1919 diary entry:

I do not want to have anything more to do with the Air Service after I get back. I shall be glad to lay down the burden I have been carrying. Now making all my plans to get away from here by 8th or 9th July. Hope to manage it, and then on home to fall at once from my high estate. It will be quite a fall, but I am not going to mind it much, provided I get off in a district somewhere and can just have enough to do—not too much.[14]

Patrick saw Pershing a few days later on 1 July to tell him that his duties were completed and that he looked forward to going home. Pershing replied that he "was very loath" to have Patrick go. Cajoling and joking with his old friend, Pershing proceeded to offer Patrick a State Department mission that would keep Patrick in Europe a further three months. Patrick declined, stating, "I am tired and too old to take on new work." He reluctantly, but rather pointedly, refused Pershing's offer.[15]

Patrick, though, was not too old or tired to make new friends, and one such newfound acquaintance proved to be of benefit to Patrick when he again headed the Air Service. Mrs. Louis (Sallie Maxwell) Bennett first wrote to General Patrick in September 1918 requesting his assistance to recover the body of her son, Louis Bennett Jr., who was killed over enemy lines while a pilot for the Royal Flying Corps.[16] Patrick recognized the family name. The Bennetts were one of the leading families of his home state of West Virginia with strong political connections, vast land and mining interests, and wealthy in the extreme. Mrs. Bennett's father-in-law served as Jefferson Davis's auditor general of Virginia and was related by marriage to "Stonewall" Jackson. Louis Bennett Sr. was a Confederate veteran (as was Mason Patrick's father), and Louis Bennett served in several political state offices before his death in 1917.

After the Armistice, General Patrick's efforts on the part of Mrs. Bennett resulted in the speedy recovery and transport of her son's remains to the States. In addition, General Patrick paved the way for Mrs. Bennett to meet with General Trenchard, and Air Ministry and Westminster Abbey officials to discuss the design, crafting, and installation of a stained glass window in Westminster Abbey in honor of American and British aviators lost in the war. Mrs. Bennett remained forever grateful to General Patrick for his intervention and kindness in these matters, and thus began a friendship that lasted for the remainder of their lives. Mrs. Bennett repaid Patrick's kindness by being, in later years, a promoter of the Air Service mission and a sounding board for Patrick's trials and tribulations while head of the Air Service. In future correspondence with Mrs. Bennett, Patrick would be uniquely forthcoming concerning his feelings and observations regarding the Air Service, personalities of the day, and politics in general.

General Patrick and Mrs. Bennett first met when Mrs. Bennett was in Paris during the months of June and July 1919 arranging for her daughter's school admission. He confided in Mrs. Bennett much of what he told General Pershing: he had had enough of the Air Service and only looked forward to retirement as an engineer. Little did either realize that in the not-too-distant future their paths would cross again because of controversy in the Air Service.

One of Patrick's last official functions was to fly to London to visit General Trenchard and the Air Ministry staff for a farewell tour and dinner. The wartime relationship that Patrick cultivated with Trenchard and the Royal Air Force would continue postwar as well. But that was yet to come. In July 1919 all Patrick looked forward to was a leisurely voyage home and to be reunited with his wife, Grace, and his adopted son, Bream.[17]

Patrick began his journey home on 13 July 1919 aboard the *Aquitania*. The ship's passengers included Secretary Crowell and his party, who proceeded to ignore Patrick for the entire crossing. Patrick described the situation thusly: "Crowell's attitude toward me is still quite peculiar. He has not indicated in any way what he intends to report—and I am not asking—nor do I give a damn."[18] Patrick was accompanied on the passage by Col. Edgar S. Gorrell, whom Patrick had entrusted in November 1918 to prepare a history and final report on U.S. air activities in Europe during the war. Patrick had much respect for Gorrell and visited his staff often as they worked diligently on the AEF Air Service history.[19] Now without benefit of an assigned aide, Patrick was generously assisted by Gorrell during the crossing to New York. Gorrell, who held several executive positions within the automotive industry after the war, continued to maintain a warm and regular correspondence with Patrick.[20]

The *Aquitania* crossing was pleasant enough, though, due to the presence on board of "a great bunch," which included Generals Hunter Liggett, Joseph Dickman, H. A. Smith, Dennis Nolan, and Harold Fiske. Billy Mitchell, who up until his transfer back to the States was Chief of the Air Service, Army of Occupation at Coblenz, went home much earlier. With Mitchell's duties at an end, Patrick sent him home, with Pershing's blessings, in late February (aboard the same ship). Mitchell made the crossing in most colorful fashion by daily executing two-hour morning inspections of the ship, complete with two buglers blasting forth to announce Mitchell's arrival as he alighted on the various decks.[21] Numerous sources note that Mitchell fully expected to be named director of military aeronautics upon his return from overseas, and Mitchell said as much aboard the *Aquitania*. But this was wishful thinking. On 8 January 1919 both Pershing and Patrick received a telegram from the War Department asking for the "return of Brigadier General William Mitchell . . . to this country immediately as his services are needed as Assistant Director Military Aeronautics." The same telegram also stated that Maj. Gen. Charles T. Menoher was appointed the director of the Air Service.[22] Whereas Mitchell did not get what he expected when he returned home, Patrick received precisely what he envisioned: command of the "not too busy" engineering district of New Orleans with his stars replaced by the eagles of his permanent colonel's rank.

When Patrick stepped off the *Aquitania*'s gangplank in New York harbor in July 1919, he desperately wanted to put the Air Service behind him. But he had to face a few hurdles before enjoying an unencumbered and leisurely slide into retirement. He was met with orders demoting him back to colonel and ordered to Washington where he was summoned before the Frear Congressional Investigation Committee. The Frear Committee was looking into aviation matters during the war years, and Patrick was singled out for the role he played in "burning" nearly 2,300 planes at war's end.[23] Patrick, in his testimony, in typical fashion assumed entire responsibility for the episode. He had received orders from the War Department to close out and satisfy all contracts and liquidate his arsenal of planes in the most efficient manner. Although much aviation-related material held by the AEF was sold to allies, neither the British nor the French wanted to purchase any of the AEF's obsolete aircraft. Shipping the aircraft back home was prohibitively expensive, not to mention that there were hundreds of new aircraft in storage at stateside bases. Patrick was left with few options. He ordered that all salvageable items (engines, armament, and instrumentation) be stripped from all aircraft, crated, and shipped back to the States. Whatever scrap metal remained was sold at auction and the remaining dope-covered fabric and wooden frame was put to the torch. In the end, each plane was surveyed by at least three salvage boards. Given the facts in a forthright manner, the committee dropped the issue.[24] Later, in a letter to a friend, he characterized the committee as being "bent on making political capital rather than getting the facts."²⁵ After General Pershing's triumphant return to the States in October, he gave testimony to the same committee. Following Pershing's testimony, Patrick wrote a letter to Pershing providing a detailed analysis regarding the future of military and commercial aviation in the United States, in the event Pershing was recalled to testify. In the letter Patrick characterized the creation of a separate Department of Aeronautics as "an organized assault upon the United States Treasury."²⁶

Patrick's appearance before the Frear Committee was his last official association regarding Air Service matters until he was called back by Pershing in 1921. But it was not long after Patrick returned to the Air Service that it became obvious he no longer thought of Air Service independence in terms of a conspiratorial raid on the Treasury. Many things happened during his absence that cumulatively and dramatically changed Patrick's perspective.

5. The First Round in the Postwar Fight for Air Service Independence, 1919–1921

Demobilization

At the end of the war, virtually all could agree that the defense organization of the United States required an air arm, but few agreed on what form it should take. The evolutionary aviation groupings adopted by the Air Service, AEF, over the Western front varied according to mission: from singular reconnaissance sorties to the massing of aircraft to achieve local air superiority, to the bombing and strafing runs that were the earliest forms of aerial interdiction. The strategic bombing concept was envisioned and attempted but remained an illusive and theoretical goal for future bomber designs. The postwar fact was that the air arm was still a lady-in-waiting for the queen of battle. Given its limited equipment, constrained training, field organization, and deployment, the Air Service role could be but little else other than direct support of the battlefield army, especially in light of the drastic draw down after the war. This was made eminently clear when a two-star artillery officer was placed in command of the Air Service.

Maj. Gen. Charles T. Menoher, who distinguished himself as commander of the 42nd (Rainbow) Division and the VI Army Corps, had absolutely no practical experience with aviation.[1] Indeed, even Billy Mitchell, assigned as Menoher's deputy, had much less practical flying experience than many of the men he commanded. Menoher was rewarded for a job well done commanding the men in the trenches, and Mitchell was rewarded for supporting the men in the trenches. Realistically, though, based on the "team player" concept that characterized Pershing's way of doing business, Mitchell was never a serious contender for the top job in the Air Service.[2] Patrick, while Pershing's friend and confidante, was not in a combat leadership role, something Pershing considered very important when the postwar commands were being considered. There were plenty of candidates to fill those command slots. By his own admission, Patrick conceded to a friend that he had been "given a chance to remain in the Air Service in a minor capacity, but . . . I treated it with spurn."[3]

In the dramatic swirl of a rapid postwar demobilization, the Air Service was caught in a slow tailspin for the next three years. This is not surprising when one examines the actions of the man who was figuratively in the pilot's seat. General Menoher was not coy about his aeronautic philosophy: the Air Service belonged—wing, stick, and propeller—to the U.S. Army. Its purpose was to give direct support to the soldier on the ground. Menoher would have none of this theoretical nonsense concerning strategic bombing or, worse still, the heretical notion that the Air Service should be independent from Army control, a force unto itself, with its own budget and supporting infrastructure. He was decidedly in favor of keeping the Air Service as a dedicated combat arm of the Army.

But Menoher need not have worried about the Air Service going its own way; the institutional weight of the executive branch of the government, the secretary of war, and the General Staff in particular would not allow it. In support of this, Secretary of War Newton D. Baker refused to endorse the Crowell report and only reluctantly made it public. Baker minced no words when he criticized the board for exceeding its mandate when it recommended an independent federalized Air Service composed of all aviation elements, military and civilian. In addition, in the midst of a rapid demobilization, the legal status of the Air Service was also tenuous because it had no permanent legal authorization to exist even in its present form. The separation of the Air Service from the Signal Corps that Pershing had initiated and President Wilson had legalized by executive order in 1918 was due to expire on 11 May 1919.[4] The Air Service was not entirely friendless though. Its congressional advocates succeeded in initiating, and Congress passing, legislation that funded the air arm for an additional year.[5]

Legislation

From the end of the war up to 1926, when the Air Corps Act was finally passed, the issue of an independent department of aeronautics never went away. It would not die, nor would it yield an organization that satisfied either of the two antagonistic camps: the entrenched Washington bureaucracy at odds with a vocal cadre of men who served in the Aviation Service on the Western front. A *Washington Star* newspaper editorial of 13 August 1919 described the direction of the future debate: "The time is coming when the theoretical views of the officers who were unfortunately compelled to stay on this side of the ocean will conflict strikingly with the practical views of the [air]men [who served] in France."[6] But this editorial was a bit tardy in its forecast, for there already existed an alarming difference of opinion among various members of the Air Service, and especially between Air Service and War Department staff members. Had the author of the editorial been privy to an exchange of memos a few months prior between Billy Mitchell and Menoher's chief of staff, Col. Oscar Westover, an Air Service pilot who was adamantly opposed to independence, he would be aware of this oft-times heated debate.[7]

There were, though, additional issues at play other than the specters of rapid demobilization, institutional rigidity and major doctrinal dogfights. Money and politics, as always, were part of the aviation equation.

In the elections of 1918, a Republican Congress was returned to Washington, a Congress that was searching for any flaws in the current administration. To prove obstinacy and shortsightedness on the part of the Democratic administration regarding Air Service independence, and/or a reorganization of all air activities, was part of the Republican agenda. It was no accident that eight separate bills introduced during the 1919 and 1920 congressional sessions (65th and 66th Congresses) sought to create a separate Department of Aeronautics.[8] It was also no coincidence that all eight bills were introduced by Republican lawmakers.

Of the eight, only two were given serious consideration, both introduced in late July 1919. The bill from Rep. Charles Curry of California was much along the same lines as what was recommended by the Crowell Commission report, calling for an executive department of aeronautics responsible for all matters pertaining to aviation.[9] The bill from Sen. Harry S. New of Indiana was very similar to the Curry proposal but much less detailed. With so much legislative flurry on the Hill regarding "independence for aviation," Secretary of War Baker lost little time convening a board of Army officers headed by General Menoher, with four senior artillery officers rounding out the obviously biased and not-so-representative membership, on 8 August to render a report on the independence issue. On the face of it, one could say that the Menoher Board made an exhaustive effort to examine thoroughly the merits of the New and Curry bills. Such was not the case. From the start the cards were stacked against an honest appraisal. To begin with, Menoher's sentiments regarding aviation independence were well known. To compound the matter, the rest of the board consisted of ground officers—artillery officers, no less—all having World War I experience, who were from the outset convinced of the paramount role that aircraft must play in the direct support of ground forces.

Menoher did not submit his report until the end of October. In that ten-week period the board examined a wealth of information "from individuals, boards, commissions, and other sources bearing upon the subject under consideration." What was unique about the board, and a stroke of surprising genius on the part of Menoher, was the initiative to request, via telegraph, the views of some fifty key World

War I division, corps, and Army commanders who had operational experience with air units within their commands.[10] As one can imagine, these telegraphic reports were, without exception, in favor of Army control of aviation, as it was during the war when AEF Air Service units were dedicated to the corps and Army commanders for operational control. Given the intensity of the congressional effort to pry aviation loose from the Army's operational control, it was no wonder that Pershing (with his keen political insight) saw the need to undermine the arguments of the independent aviation advocates by convening the Dickman and Foulois Boards, both arguing squarely in favor of Army control. One must remember that Pershing, as the commander of the Mexican Punitive Expedition in 1916 and again as the AEF commander, witnessed firsthand the necessity for a viable aviation arm, but as a combat arm dedicated first and foremost to ground support.

After his triumphal return from Europe in September 1919, Pershing left little doubt about his convictions regarding military aviation. But this particular congressional testimony was a small part of his views within the much larger question Congress was considering: the Army Reorganization (Baker/March) bill. On 2 November, during his three-day testimony before the House and Senate Military Affairs Committees, Pershing reiterated the views held in the Dickman and Menoher reports that the Air Service for military purposes should remain part of the Army.[11] Pershing was even more explicit in his views when Menoher wrote to him on 16 December requesting a clarification of Pershing's congressional testimony regarding a separate Air Service.[12] Menoher opined: "I believe that the press of the country either misunderstands or misinterprets your views on the needs of the Air Service . . . leading the [public] to believe that you favor such a separate Air Service as is proposed in the New and Curry bills."[13] Pershing's exasperated response left no room for doubt:

I am at a loss to understand how my opinion on the question of a separate Air Service . . . could be misinterpreted. In those hearings, and on many other occasions, I expressed my view that the Air Service for military purposes should remain a part of the Army . . . coordinate with the Infantry, Cavalry and Artillery [and] an air force should not be established as a combatant force distinct from the Army and the Navy.[14]

Pershing was not alone in this view. Gen. George Bell Jr., who commanded the 33rd Division during the Meuse-Argonne campaign, was direct when he pointedly complained about the behavior of Air Service personnel:

No people in this war needed discipline more than the aviators and none had less. All the attention was given to handling the machines and but little thought was had of discipline.

The result was more or less of a mob with great loss of efficiency, as strict discipline is the foundation stone of military success.[15]

Addressing the independence issue, even the most sympathetic response, from Col. Dennis Nolan, the chief of AEF Intelligence, was qualified: "Aerial bombing against such objectives as towns, railroads, bridges, etc., produce little material effect; such operations are certainly the only ones that could be carried out by a force acting independent of the commander in the field."[16] What is interesting about the majority of responses to the Menoher Board, besides the unanimous view against Air Service independence, were the repetitive themes concerning "discipline" and "control." For example, General Dickman: "The factor of greatest importance in aeronautical service is discipline."[17] Mitchell's and Foulois's antics prior to and just after Patrick's first appointment as Air Service chief obviously soured many ground commanders on the notion of any degree of Air Service autonomy. The views of many independence-minded Air Service officers ran counter to Pershing's belief that "the Infantry still remains the backbone of the attack, and the role of the other arms is to help it reach the enemy." As noted previously, Mitchell made enemies on the Western front, and they were not only confined to the hated hun.

Although Mitchell was ordered to return to Washington as the assistant director of military aeronautics after the war, he returned to find the position abolished in the postwar reorganization. In March 1919 he was assigned to Menoher's staff as G-3 (training and operations officer). In the postwar military aviation controversy Mitchell had no allies within the War Department, but this did not prevent him from pursuing his personal agenda.

The first postwar public spotlight, though, compliments of the New and Curry bill hearings, had to be shared with one of Mitchell's bitter antagonists, Major Benny Foulois. Two weeks prior to Pershing's testimony, Mitchell and Foulois had offered their own views on the future of aviation. In a rather stunning turn of events Mitchell was a milquetoast while Foulois was a firebrand. Mitchell insisted that his campaign be handled "on a broad basis and a high plane."[18] Foulois, by comparison, was engaged in a one-man dogfight with the General Staff. In his October 1919 testimony before the Joint Committee on Military Affairs, in the midst of the debate surrounding the New and Curry bills and the War Department's Army Reorganization proposal (House Bill S.2715), Foulois was at his combative best:

[I]n my opinion, the War Department has earned no right or title to claim further control over aviation or the aircraft industry of the United States. . . . The War Department through its policy-making body, the General Staff of the Army, is *primarily responsible* for the *present unsatisfactory, disorganized, and most critical situation which now exists in all aviation matters throughout the United States.*[19]

A distinction is necessary here regarding the ways in which Mitchell and Foulois justified the case for Air Service independence. Mitchell's rationale for aviation independence was based on justifying a separate mission that only the Air Service could execute, while Foulois's justification was based on the War Department's "inability to adequately provide for and direct the Army's air arm."[20] This distinction is critically important when one seeks to understand this particular era because it underlies the issue of Billy Mitchell's ultimate fate and, by comparison, how individuals like Mason Patrick and Benny Foulois fared when taking on the War Department. Whereas Foulois and, to a much lesser extent, Patrick could be critical of the War Department and General Staff's disposition toward the Air Service, neither of them went so far as to call for the displacement of the infantry with the air force as the decisive element of warfare, as did Mitchell. Such a notion was sacrilege. It was this emotional issue, and to a lesser extent the issue of Air Service independence, which made Mitchell a pariah in the eyes of the War Department and the General Staff. In particular, the members of the General Staff, many combat veterans of World War I, found Mitchell's notions to be at best flights of fancy, and at worst heretical ideas that insulted the memory of the AEF's effort on the Western front. In addition, Mitchell had no Air Service allies on the General Staff: of eighty-four members, not one was from the Air Service.[21] To make matters worse, Mitchell further alienated himself from Secretary of War Baker by writing in April 1919 that the concept of total war, directly involving the civilian populations of the belligerents, was justified.[22] Secretary Baker was opposed to such a concept. His admonition concerning such was forwarded to Chief of Staff March, who in turn telegraphed Patrick, who in turn forwarded an 11 November 1918 letter to Pershing discussing same. Baker thought that an independent air arm would lead to the targeting of enemy cities, as was the case with Britain's Independent Bombing Force."[23] To compound the problem even more, in his quest to justify a mission unique to the Air Service, Mitchell would make the Navy Department a most dire enemy through his 1919–20 legislative efforts and their aftermath.

During this period, a large portion of Patrick's hiatus from the Air Service, the legislative Air Service battle was fought. Of the eight bills introduced during this period to reorganize the Air Service, only one (News's) was even reported out of committee. Secretary Baker proved to be an astute political in-fighter: quickly grasping the possible effects of the Crowell Commission's report, Baker convened the Menoher Board, the results of which, combined with the Dickman Board report, all but buried any effect the Crowell report would possibly have on pending "independent leaning" committee bills. Ultimately, it was the War Department's legislative initiative that became law in the form of the Army Reorganization Act of 4 June 1920. The Air Service was given statutory recognition and made one of the combatant arms of the Army. It was a relationship precisely as described by Pershing in his January 1920 letter to General Menoher. The General Staff still reigned supreme.

Given the drastic cuts in manpower, however, the General Staff ruled over an anemic organization. Whereas the June 1920 Army Reorganization Act allowed for 1,516 officers and 16,000 enlisted, by 1922 there were 950 officers (of which only 150 had Regular appointments) and 9,000 enlisted men in the flying arm.[24]

The original legislation that the General Staff supported in the Army Reorganization bill debates was rather harsh on the Air Service. Had the original General Staff–supported provisions passed, it would have been problematic whether the Air Service would have survived. Pershing's direct intervention in support of the legislation making the Air Service a combatant arm of the Army would formalize what he had done in 1917 in the AEF. In the end, the Army Reorganization Act basically preserved the status quo between the General Staff and the Air Service and thus can be characterized as a resounding defeat for those wanting independence. But there were distinct benefits as well:

1. A slight increase in personnel was authorized;
2. Provisions for flying pay were expanded, as well as providing that not more than 10 percent of the officers of each individual grade below that of brigadier general be nonfliers;
3. The Air Service was designated as a unique supply branch charged with the development, procurement, storage, and issue of its aeronautical related material; and
4. All tactical flying units were to be commanded by flying officers.[25]

While some advantages were gained for the Air Service in this reorganization, Mitchell's 1919–20 push to secure legislation for a centralized aeronautics organization also seriously hampered Navy efforts to secure an effective reorganization plan for their own air arm. After Mitchell's independence legislation went down to defeat, he was instrumental in having a rider attached to an Army appropriations bill that gave the Army Air Service control over naval air stations. With last-minute intervention, the Navy was successful in retaining control of its air bases but limited by the final legislation to only six.[26] The big naval guns began to take sight on Mitchell.

Reorganization

Without exception, the literature of this period characterizes the 1920 Army Reorganization Act as "a resounding defeat" or that "the General Staff . . . and the War Department had won a veritable triumph" over the Air Service.[27] This characterization, though, is only applicable when considering the issue of independence. It is a misrepresentation to portray the 1920 Army Reorganization Act as being wholly without benefit to the Air Service, citing only the fact that independence was not realized. It is more appropriate to characterize the end result as a much welcome (and surprising) incremental victory, as Patrick did, given the tenor of the times. Considering alone the overriding issue of economy in government, there was very little chance that a new aviation bureaucracy would have been sanctioned. Add to this the historic desire for a small standing army, the outrageous aviation expenditures (and associated fraud) during the war, and the isolationist tendencies of the populace, it is no wonder that the independence movement was soundly defeated.

Rays of hope, though, did exist. Even Secretary of War Baker, who did not approve of an independent aviation arm, was quite sympathetic and conversant regarding the current status and future promise of military aviation. He addressed this issue at length in his 1919 annual report, even going so far as to state that it was "perhaps dangerous to attempt any limitation upon the future [of aviation] based even upon the most favorable view of present attainment."[28] Add to this the fact that the gains for the Air Service in the 1920 legislation, though modest, were very positive steps on the long road to independence. The passage of the 1920 Army Reorganization Act is not only notable for the modest gains it allowed the new Army combat arm, but it can also be characterized as a historic fork in the road offering two approaches to Air Service independence: the radical trajectory Mitchell and his disciples had embarked on, and the gradualist, within the system path, represented by Patrick and Foulois.

As direct and confrontational as Foulois could be at times in his congressional testimony, one would think that he and Mitchell were in collusion in a campaign for Air Service independence. Far from it. Their mutual antipathy, begun during the war and growing more intense since the Armistice, manifested itself in their daily deportment. Even though they had adjacent offices, they ignored one another.[29]

Things had not changed much since the war, both from the way certain Air Service officers interacted and from the way Air Service officers as a whole related to the War Department, the General Staff, and the Navy. While General Menoher and his chief of staff, Colonel Westover, an aviator, were firmly against an independent Air Service, many below that level held just the opposite view.[30]

After the passage of the 1920 Reorganization Act, which Mitchell described as a "crushing defeat," he and his disciples changed tactics. Mitchell had been comparatively mild in his criticism of the War Department and the General Staff, especially when compared to Benny Foulois, but he became much more strident and prolific. While most of his written work on aviation prior to 1921 was for an "in-house" audience, being published in *U.S. Air Service,* those items published in 1921, and especially from 1925 on, were for the public in general, appearing in *Review of Reviews,* the National Geographic, Popular Science Monthly, and the *Saturday Evening Post.* His first book, *Our Air Force* (1921), emphasized the concept of Mitchell's cherished Department of Aeronautics. Subtitled *The Keystone of National Defense,* Mitchell was adamant in calling for aviation independence.[31] It was his plan to take his case directly "to the President and the American people to hear our cause."[32] Just like a politician on the stump, Mitchell made every effort to make his case heard.

Mitchell, as the son of a one-time U.S. senator, knew a bit about politics, and money was a most important ingredient in the political process. In pushing strenuously for an independent air arm, Mitchell wanted to undercut one of the main objections to independence, that being the cost involved to underwrite the bureaucracy of a new military department. If, in this time of austerity and "economy in government," Congress could not be convinced to commit additional funding for an independent air force, then perhaps Congress could be persuaded that the current priorities of the War Department budget needed revision.

The Sinking of Ships

Mitchell actually began his effort of revising the War Department budget in February 1920 when he appeared before the House Committee on Military Affairs and "transfixed" the committee's members as he described how American land-based aviation would savage an invading enemy fleet. Mitchell went on to note casually that a battleship cost the equivalent of 1,000 bombers and lamented the investment in what would soon prove to be a "worthless" weapon.[33] Thus, Mitchell's entre into the realm of coastal defense was his attempt to make air power a prominent, if not dominant, arbiter of that mission. Hand in hand with this approach was his desire to take over the responsibility for naval air as well. But to lay the groundwork to justify such ambitious plans Mitchell needed to prove that bombers could indeed sink and therefore literally displace battleships as the first line of coastal defense. Mitchell's next step was to call for a contest between bombers and battleships to prove that theory.

The Navy had already planned for bombing tests using solely naval aviation assets. Mitchell, though, now waged a year-long campaign to pit Air Service bombers against a battleship. During this period Mitchell immersed himself in a public and congressional relations campaign to win approval of the idea. Ever the dynamic publicist, he was ultimately successful in this endeavor and bloodied the Navy even more grievously than during his 1919–20 legislative battle. But Mitchell made lasting enemies along the way, adversaries who (in addition to General Staff and War Department leadership) would contribute to his final demise.

Some of those antagonists lashed out at Mitchell in May 1921. In the aftermath of the crash of a Navy flight that resulted in the loss of seven men, Mitchell publicly called for the centralized control of aeronautics. What he viewed as an opportunity to press his case for aeronautical independence was, for many others, a most intemperate outburst. General Menoher personally denounced him. The Navy, in the person of Adm. William A. Moffett, excoriated Mitchell's self-serving criticism. Moffett, who would soon take over the newly established Bureau of Naval Aeronautics, charged that Mitchell "used the recent disaster . . . as an argument in favor of a united air service. This disaster had nothing whatever to do with the united air service [issue]."[34]

Having successfully alienated Menoher, the senior leadership of the War and Navy Departments, and the General Staff, Mitchell had only one place to turn: by default, public support was his only option. It was to the public alone that he owed his continued place in the Air Service hierarchy. As such, he had cause for concern. Mitchell's ill-considered remarks concerning the fatal naval aircraft crash, combined with his many run-ins with the Navy and his frequent clashes with his direct superior, prompted Menoher to ask Secretary of War John W. Weeks to dismiss his assistant chief in mid-June 1921. This occurred just a week prior to the scheduled contest between bomber and battleship.

When the Menoher/Mitchell imbroglio hit the press, Secretary Weeks offered his opinion to the *New York Sun* "that Army discipline and precedent would probably dictate that he side with Menoher and move Mitchell to another post."[35] The public would have none of this and the Harding administration took notice. Weeks gave Mitchell a verbal and written reprimand stressing that henceforth he "pull with the team." Menoher withdrew his request that Mitchell be sacked and Mitchell's make-or-break public affairs extravaganza against the battleship Navy took place as scheduled. The *Ostfriesland* and several other ships were sent to the bottom almost effortlessly by the planes of Mitchell's Langley Airfield–based First Provisional Air Brigade, aided by the fact that Mitchell brazenly broke some of the more restrictive of the numerous Navy ground rules of the test in the process. Mitchell's dramatic success ensured his continued place in the Air Service if not his elevation to the top job, as many newspaper editorials demanded.[36]

Not only did many of the nation's major newspapers glowingly commend Mitchell's accomplishment, but several of the more notable editorials went further, much to the chagrin of the Navy. The *New York Times* gave voice to what the Navy tried most desperately to squelch: "Brigadier General William Mitchell's dictum that the 'air force will constitute the first line of defense of the country' . . . no longer seems fanciful to open-minded champions of the capital ship."[37] Mitchell had hit on the perfect flanking maneuver in his war with the U.S. Navy. If his engagement resulting in the sinking of the *Ostfriesland* can be termed a decisive tactical victory, its aftermath could very well be described as a potential strategic nightmare for the blue-water Navy. A combination of economy in government and moral-based disarmament movements threatened the Navy with the possible loss of its reason for being. The stakes were very high indeed, and Mitchell successfully called the Navy's bluff. What raised the stakes even higher was disarmament advocate (and influential) Sen. William E. Borah's declaration "that the battleship is practically obsolete."[38] On the Senate floor Borah questioned the need for spending $240 million for six new *Indiana*-class battleships that were currently under construction. Borah asked why these funds should be expended if "with sufficient airplane and submarine protection this country was perfectly safe from attack"?[39]

Menoher Resigns

A fortnight before, as part of his justification for Mitchell's removal, Air Service Chief Menoher complained to Secretary Weeks that Mitchell had been pushing him out of the Air Service picture entirely, by "[enhancing] his own prestige at the expense of and to the detriment of the prestige of his immediate commanding officer."[40] With the *Ostfriesland* sinking, Mitchell was the object of news stories around the world. And even though Menoher went out of his way to praise Mitchell's accomplishment, Mitchell belittled the gentlemanly peace offering by privately criticizing what he saw as Menoher's failure to grasp the significance of the event.[41]

The confluence of these interrelated events ultimately brought the Mitchell/Navy confrontation to a head and resulted in Patrick's replacement of Menoher as Air Service chief. Both sides stood to lose much indeed. Mitchell thought that his spectacular showing against the battleship admirals was proof of the aircraft's superiority. The Navy was threatened not only by Mitchell but by the Conference for the Limitation of Naval Armaments soon to take place in Washington. The U.S. Navy needed a victory. The only authorized public report of the bombing tests was issued on 20 August 1921 by the Joint Army and Navy Board, whose members included the senior officers of their respective services. The report, signed by Gen. John J. Pershing, the senior member of the board, decreed that, although the airplane posed a serious threat to any ship afloat, the battleship was "still the backbone of the fleet and the bulwark of the nation's sea defense and will so remain so long as safe navigation of the sea for purposes of trade or transportation is vital to success in war."[42] In addition, to counter this threat, improvements in battleship design and the construction of aircraft carriers were recommended. The battleship sailed on.

Mitchell was incredulous and even more determined to make his case. To do so, he wrote a detailed report on the results and future implications of the bombing experiments held thus far.[43] Submitted to Menoher, it was a scathing point-by-point rebuttal of the Joint Board report, a detailed review of what the bombing tests proved thus far, and it ended with a call for an independent air force and the establishment of a department of national defense. To Menoher this was just another instance of Mitchell being Mitchell, and it is probable that Menoher's intention was to either pigeonhole the volatile report or, more likely, forward it to Secretary of War Weeks as one more example of Mitchell's wayward character.[44] But Secretary Weeks did not get the chance to read Mitchell's report first-hand. He read about it in the *New York Times* of 14 September.[45] Menoher was apoplectic. It is not clear how the report (which Mitchell was under clear orders not to release outside the War Department) found its way to the press. But, upon reflection, it is indeed a testament to Mitchell's popularity when one realizes that when Mitchell "leaked" his rebuttal report on the bombing trials he was directly challenging the very popular General Pershing, whose signature alone graced the Joint Board report. Hap Arnold himself commented on this turn of events many years later when he noted, "If he [Mitchell] could attack the signature of the Chief of Staff who was General Pershing at that, it was plain it was going to take a lot to stop Billy Mitchell."[46] Whether he or his activist public affairs apparatus released it is of no consequence; only the result is important. It was at this point that Menoher requested Weeks's backing to discipline Mitchell, and if that support was not forthcoming Menoher's resignation would be. Menoher's resignation letter was accepted by Weeks on 15 September 1921.[47]

Menoher's removal can be best attributed to two factors, the Mitchell issue being primary. A closely related problem, though, dealt with Menoher's overall competency, evidenced by the way in which he approached his duties. Six weeks before Patrick arrived, Menoher had written to the War Department requesting that certain personnel problems peculiar to the Air Service be considered in congressional hearings on pending legislation. The Adjutant General's response was quick and severe: "Your letter was not received until the hearings on the bill had closed. In fact, your letter was not even written until after the hearings on this bill had been closed."[48] By mid-August Menoher's grip on the Air Service's top job was coming swiftly to an end.

Menoher's removal was far and away a byproduct of the political fallout that would have ensued had Mitchell been ousted, especially in the afterglow of the

Ostfriesland coup. Although there has been much conjecture as to precisely why Weeks chose to not discipline Mitchell and thereby invite, if not guarantee, Menoher's resignation, a close look at the timing of the affair is warranted. Secretary of War John W. Weeks, who had only taken over from Baker several months before, had experienced the brunt of Menoher's inability to control Mitchell in the run-up to the bombing tests. In addition, the political idea of using the airplane as an "economy" plank of the budget had appeal, especially given the hugely popular reception given Mitchell. In addition, and most important, thanks to the new chief of staff General John J. Pershing, Weeks had someone waiting in the wings. Mason Patrick was conveniently located at Camp Humphreys, a half-hour drive south of Washington.[49]

Pershing began working the resolution of the Menoher-Mitchell issue soon after taking over as chief of staff on 1 July 1921. The preparations for Menoher's replacement began in earnest in late August, a full three weeks prior to the leaked Mitchell report that led to Menoher's ultimatum to Weeks. On 29 August Secretary Weeks had visited Patrick at Camp Humphreys, ostensibly to "look over the post." In reality, as Pershing later confided to Patrick, "Mr. Weeks came down to look over you. That means something about the Air Service."[50] In fact, Patrick's appointment was taken as a foregone conclusion by September in senior Army circles and the news generated very positive agreement. Maj. Gen. H. F. Hodges, Third Corps Area Commander based at Fort Howard, Maryland, wrote to Patrick on 22 September 1921: "I . . . congratulate you and the Army at large most sincerely. I have discussed with others the subject of Menoher's successor, and in every case there has been unequivocal and enthusiastic support for you."[51] Without exception, the Army's flag-rank fraternity welcomed Patrick's appointment and forwarded their enthusiastic congratulations.[52] Even Brig. Gen. Hugh A. Drum, at that time the assistant commandant of the General Service Schools at Fort Leavenworth and not inclined to support the Air Service, sent effusive congratulations.[53]

Patrick had distinct advantages that Menoher and Mitchell did not. "Most members of the General Staff liked Patrick; he was, after all, one of their own."[54] The fact that Patrick headed the Air Service during the late war gave him a credibility among the Air Service troops that Menoher distinctly lacked. Not that everyone endorsed Patrick's appointment. Eddie Rickenbacker, noted American air ace and close friend to Mitchell, opined, "General Patrick is a capable soldier but he knows nothing of the Air Service." Rickenbacker summed up with his comment that "the appointment is as sensible as making General Pershing admiral of the Swiss Navy."[55] Rickenbacker's observation may have made for colorful commentary in the popular press, but the crucial support for Patrick, the Army's senior leadership, was solid. In addition, many Air Service officers approved of Patrick.

Among these was Henry "Hap" Arnold, who noted, "the new Chief's experience with air power was a secondary consideration. In the eyes of the General Staff, it was the experience with Gen. Mitchell that counted. If there was any Officer in the Army who should be able to control him, Patrick was the man."[56]

Patrick was fortunate in this regard, for he would soon require all the help he could get. Once again, Pershing requested help from his West Point classmate for what seemed an intractable problem. It was a predicament that Patrick had solved before and, Pershing hoped, he would solve again.

6. "To Command in Fact as Well as in Name"

The Path Back to Air Service Command

The replacement of Menoher with Patrick was singularly convenient. After spending the last three months of 1919 in a cramped walk-up apartment in New Orleans while supervising the engineering district there, Mason Patrick was recalled to Engineering Headquarters in Washington in January 1920 with a swiftness unusual for the postwar army. Patrick had foreknowledge of this move: when he stopped in Washington the previous September he was told he might not be in New Orleans for very long.[1] This may have prompted Patrick to speculate on his future, especially with regard to the Air Service, when he wrote to his friend, Mrs. Bennett. When she mentioned the possibility that he might again head the Air Service, Patrick did not reject the possibility outright:

[T]here is too much pulling and hauling in that Service just now. They are playing politics a bit, and they do not know, some of them, what they do want. It is better to be out of it for the present. Gen. Mitchell was one of my subordinates in France. He was in command of the Air Forces at the front, and in some ways he did quite well. He is enthusiastic, but erratic and not well balanced. I left a pretty clean slate behind me, I think. I have the consciousness of having worked pretty hard, of having done the best I could, and this is enough for me now.[2]

Back in Washington, Patrick performed two months of "special duty" as "Special Assistant to the Chief of Engineers" which consisted of drawing up contingency deployment plans for engineering personnel. He was then retained at headquarters on an ad hoc basis for over a year doing little substantive work. As he described it, he was moved with an unexpected "suddenness . . . for an indefinite period . . . for unspecified duty."³ His move was made permanent in June 1921 with an assignment to Camp Humphreys (present-day Fort Belvoir), twenty miles south of Washington.

Within the bucolic environment of Camp Humphreys, Patrick was commander of the engineering school and in an ideal setting to enjoy "the outdoor life . . . with leisure time occupied by riding, fishing and hunting."⁴ He wanted to remain in this position until retirement, but the chaos in which the Air Service found itself put an early end to his desire for an uneventful and provincial preretirement tour.

The radical restructuring of the Air Service in manpower and funding cuts brought the organization to the brink of collapse. It was neither efficient nor able to address its peacetime (let alone its wartime) mission. As a fledgling combat arm the Air Service, under the influence of a revolutionary (in the eyes of the General Staff) Billy Mitchell, did not fit the team-player image so revered by Pershing. In addition, Mitchell had antagonized the senior leadership of both the Army and the Navy. The Air Services' role in national defense was so wrought with emotional turmoil that a dispassionate examination of the issue was all but impossible.

What Mitchell had succeeded in doing was to move outside the accepted bounds of War Department political behavior to publicize his cause.⁵ He had been so successful at this endeavor as to stop his dismissal by Secretary of War Weeks. If Secretary Weeks had moved against Mitchell, the public and political repercussions would have been unavoidable.

History was repeating itself. Again Pershing had a fiasco within his air arm, and there was little officer depth to choose from to replace the offending personnel. And, yet again, Pershing was, doubtless, of two minds on the issue: though a disciplinarian, he admired courage under fire and effectiveness in combat. Mitchell displayed much of this mettle, but did so with a streak of independence that ultimately destroyed Menoher's authority. Pershing went with a proven winner. The spirit of Mason Patrick was still going strong and, remarkably for a fifty-eight-year-old soldier, so was the flesh.

When the formal offer was made to Patrick on 28 September 1921 to once again take over the Air Service, Patrick's flash of spirit was obvious. He was called to the office of Maj. Gen. James G. Harbord, the deputy chief of staff, and was told that Secretary Weeks had recommended to President Harding that Patrick replace Menoher. Patrick's reply was to the point: "I was very loathe to accept it [and] not at all certain that I would be willing to take upon myself what I could plainly foresee would be a most onerous duty."⁶ Harbord prevailed upon Patrick to at least think it over. Secretary Weeks, though, did not give Patrick much time to consider the offer. President Harding sent the nomination to the Senate the next day and two days later the Senate confirmed the appointment. Patrick accepted the position effective 5 October 1921 and was on duty at his desk at Air Service headquarters on Friday, 7 October.[7]

The duties that Patrick assumed included, of course, command over his deputy, Billy Mitchell, and that is precisely what Patrick intended to do. Both Pershing and Patrick were well aware of the positive and negative factors regarding the Mitchell issue. Pershing discussed these with Patrick, much as they had discussed the same issues during the critical AEF days.[8] Patrick's orders certainly did not include putting an end to Mitchell's career or stifling his creative genius. Patrick thought too highly of Mitchell to take that approach. He described Mitchell thusly:

Mitchell is very likable and has ability; his ego is highly developed and he has an undoubted love for the limelight, a desire to be in the public eye. He is forceful, aggressive, spectacular. He had a better knowledge of the tactics of air fighting than any man in this country . . . and would lose no opportunity to take a fling at the Navy. I think I understood quite well his characteristics, the good in him—and there was much of it—and his faults.[9]

On his first day in the office Patrick called the entire staff in for a general discussion of Air Service matters.[10] At the meeting's conclusion, he asked Mitchell to remain. During the ensuing discussion Mitchell commented that in anticipation of Patrick's arrival he had taken the liberty of preparing a new organizational plan for the service and asked for Patrick's review. Mitchell delivered the plan to Patrick that afternoon. Patrick noted that even the most casual reading of the plan revealed that if it were put into operation Mitchell "would practically have charge of most of the Air Service activities while the Chief of the Air Service would have but little control over him." The next day Patrick returned the plan "disapproved" and told Mitchell: "I [will] be Chief of the Air Service in fact as well as in name." In addition, Patrick told Mitchell to give no order without his approval and all decisions "in every case" would be made by Patrick. Mitchell threatened resignation. Patrick was nonplussed and even escorted Mitchell to General Harbord's office so that Mitchell could formally tender his resignation. Harbord's absence from the office, as it was Saturday, allowed Mitchell a grace period to think over his threat. On Monday, 10 October, Patrick, with Mitchell in tow, returned to Harbord's office and explained Saturday's turn of events. Harbord immediately offered to accept Mitchell's resignation. As described by Hap Arnold many years later, Mitchell "backed down and agreed to Patrick's terms."[11] Prior to the meeting, Patrick had already put those terms in writing and reviewed them with Harbord, who asked if Mitchell would abide by them.[12] What is noteworthy, and up until now overlooked, is that Harbord had written to Menoher on 17 September and requested "the difference between the duties of the Chief of Air Service and the Assistant Chief of Air Service."[13] In essence, Harbord, probably at Pershing's direction, had prepared for a Mitchell power-play.

Ultimately, Mitchell agreed to a detailed delineation of the responsibilities of the assistant chief, Mitchell acknowledged that General Patrick was the final authority within the arm concerning all Air Service matters and agreed to the review of all speeches and written work destined for public release. Mitchell reiterated much the same thing in the presence of Pershing and Patrick when the chief of staff returned to his office on 17 October.[14] Indeed, the whole affair must have been very unsettling to Mitchell, for he immediately asked for permission to do an unscheduled and very leisurely inspection of Langley Field, where his Provisional Air Brigade was in the process of being disbanded and where Mitchell would be in the company of many friends.

In the short term, Patrick had immediately established his command legitimacy and demonstrated that he had both Harbord's and Pershing's support. In the longer term, Patrick was faced with controlling Mitchell and coming to grips with the current and projected condition of the Air Service. Patrick characterized his new command as being in "as chaotic a condition as I had found it when some three years before I had been placed in charge of it in France."[15]

To Fly and Fight for Funding:
Laying the Groundwork in 1922

Patrick's challenge was immense. The Air Service in his estimation was in a critical state: obsolescent aircraft, overstated claims about current (and projected) capabilities, War Department conservatism, underfunding, and promotion issues were the major problems. During his first two years in office, Patrick was concerned primarily with maintaining the Air Service as a viable organization, establishing its credibility within the War Department establishment, and making his superiors aware of the abysmal state of the Air Service. Once he was successful for the most part in this endeavor, he would then move dramatically forward to solidify the Air Services' position by seeking special status for his organization by the requisite congressional legislation. But his agenda in his initial two years centered on two halves of one basic issue: credibility. One part concerned discipline and the other concerned Air Service organization, and more specifically, how doctrine, roles, and missions were executed within that organization.

Discipline or, more to the point, getting Air Service personnel to work within the established system was Patrick's first short-term vital concern. One month after taking command, he received a congratulatory letter from Maj. Gen. Francis J. Kernan, a friend of long standing who at that time was commanding general of the Philippine Department. Kernan got straight to the point: "It's the Air Service [personnel] which gives me more trouble than all others put together."[16] Patrick addressed this issue in his reply:

It is the youth and inexperience of its officers whom it is necessary to place in responsible positions that are largely the causes of the trouble which is found. . . . I mean impressing upon these officers as firmly as may be necessary the fact that their duty must be performed properly, that the constituted authorities must exercise efficient supervision over them, and that they must learn the essentials of discipline.[17]

Discipline was not the only emphasis Patrick had in mind when he took over the Air Service, but he realized that Air Service officer discipline was the key factor in his effort to disseminate a credible Air Service message. The Air Service needed a credibility campaign, not a publicity campaign. Patrick realized the need to stress the mission of the Air Service, not only from within the War Department establishment but also from a position of strength. Acquiring that strength meant two things: (1) control over Mitchell, which Patrick amply demonstrated during the initial confrontation with his opinionated deputy, and (2) keeping Mitchell out of the popular press.

Mitchell had concluded his congressionally authorized bombing tests by the end of September 1921 and was in the process of disbanding his Provisional Air Brigade at Langley when Mason Patrick took over the Air Service in October. It did not take long for Patrick to ensure that all parties concerned would enjoy a respite from Mitchell. Beginning in December, Patrick sent him on a four-month-long inspection tour of European aviation. The official orders directed Mitchell to "obtain complete and exhaustive information" on all European aviation matters.[18] Patrick was careful to issue follow-up cautionary instructions to Mitchell as well. On 7 December 1921 Patrick directed Mitchell: "Your mission is primarily and exclusively the gathering of information and not the discussion of Air Service or aeronautical policies."[19] One Mitchell biographer infers that Mitchell was removed from Washington to avoid any embarrassing outburst during the sensitive Limitation of Armaments Conference then in session in the city.[20] This contention is undermined by the fact that General Patrick appointed Mitchell, with the approval of Secretary Weeks, as the U.S. Army's aviation expert on the conference's technical committee.

Mitchell's four-month European tour gave Patrick a lengthy period in which to come to grips with the issues confronting the Air Service.[21] Doubtless Mitchell's incessant hectoring prior to his departure about the many shortcomings of the Air Service had some effect on Patrick.[22] But Patrick was his own expert when it came to organizational administration. With the Mitchell problem on hold, or at least on tour, Patrick's next step as head of his new organization was to see exactly how the Air Service was organized. On a whirlwind tour, Patrick quickly determined that the vast majority of the twenty-two Air Service equipment depots could be eliminated. Dilapidated sheds and hangars provided scant protection for already obsolete aircraft and crated Liberty engines. He ordered seventeen depots closed, the worthless planes sold or scrapped, and the salvageable items consolidated at five remaining regional depots.[23] He later took the same approach with flight training, concentrating all pilot training in San Antonio, Texas.

In a period free of Mitchell's frenetic distractions, Patrick made good use of his time by completing his first formal written project: the Air Service annual report for 1921.[24] Even though the report technically ended on 30 June 1921 when General Menoher was still the Air Service chief, its completion in December afforded Patrick a platform to address what he saw as the Service's major deficiencies.

When Patrick submitted his first *Annual Report of the Chief of the Air Service* in 1921, his lead sentence acknowledged two critical elements of the 1920 legislation: "The Army reorganization act established the Air Service as a combat branch of the line of the Army, and invested it with the responsibility for the development and supply of its own technical equipment."[25] Patrick was proud of these accomplishments and rightfully touted them. Although he was not at this point an advocate for Air Service independence, he realized the immediate organizational benefits of the legislation, although others did not.[26]

It was readily apparent to Patrick that problems within the commissioned force must be addressed first. The authorized commissioned strength of the Air Service was fixed at 1,516 officers by the Army Reorganization Act of 1920. At the beginning of Patrick's tenure, only 975 officers were assigned; of those, 642 were pilots.[27] To further complicate the matter, the War Department wrongly interpreted a key section of the act, which permitted the automatic loss of 9 percent of the pilot/observer washouts instead of retaining them in sorely needed administrative duties.[28]

In this report Patrick also addressed the serious shortfall among the highly technical enlisted force (only 65 percent manned) and the understaffed civilian force. The civilian employees of the Office of the Chief of Air Service had been arbitrarily reduced by the Bureau of the Budget from 725 on 1 January 1921 to 301 by 30 June.[29]

Patrick emphasized the dire problems associated with personnel shortfalls in the 1921 report by touching on organizational doctrine: "A study of the tactical and strategical employment of aviation discloses two distinct classifications of military air power: air service and air force."[30] Patrick stated that a properly balanced military air organization would result in 20 percent of the total strength being devoted to defensive observation units, constituting the "air service" portion, and the remaining 80 percent devoted to an offensive combat "air force."

With severe funding shortages for personnel and new aircraft in the Air Service, the defensive portion constituted almost half of the total organization. In Patrick's view the Air Service was not properly balanced to be an effective *offensive* combat force. He addressed the issues of updating the physical infrastructure of the Air Service as well as the need for new aircraft to fulfill the offensive and defensive roles of his "air force" and "air service."

Doctrine, Roles, and Missions: Patrick's Evolutionary Change

Patrick's views in this December 1921 report differ significantly from the views he expressed at the end of the war, as well as the views expressed in his 11 November 1919 letter to Pershing after his appearance before the Frear Committee. Both the Pershing-commissioned Patrick (Foulois-chaired) and Dickman postwar review boards emphasized the overriding importance of the Air Service's support (observation) role. In addition, the May 1919 Patrick Board report confirmed the status quo: Air Service missions and units were integral parts of the corps and field army structure, and there was no mention of a "strategical" mission. Patrick's airpower philosophy of 1919 had been far different than that which he held two years later as the new Air Service chief.

The Air Service organizational and employment doctrine that Patrick espoused in his initial months as chief was a clear indicator of the direction in which he wanted to take the Air Service.[31] As an extraordinarily adept manager, he was particularly aware of the need for clearly defined roles and missions for an organization. He began to form the basis of this newly invigorated marriage of mission and organization when he made the doctrinal distinction between the defensively oriented "air service" support units of World War I vintage with the newly coined "air force" offensive units found in his 1921 Air Service annual report. With this initiative, Patrick took the first step to codify an offensive doctrine of the Air Service, separate from a strictly ground-support function.[32] It is interesting to note that, by comparison, the institutional Army accepted Mitchell's St. Mihiel air effort as a most appropriate use of the air arm, what today would be termed "close air support" and "battlefield interdiction."[33] But Patrick was moving far beyond the St. Mihiel concept to a clearly delineated acknowledgment of the critical value of an offensive air superiority combat force. Just like Mitchell, Patrick was bucking the system. But Patrick sought to change it from within by spoonfeeding his desired revisions not to the General Staff but to the secretary of war and Congress.

Prior to Patrick's arrival, what the airmen proposed and what the General Staff promulgated as official air doctrine were very much at odds. Patrick was well aware of this disparity. Upon assuming his position as Air Service chief, he noted, "[The Air Service] not only had to find itself, but it had to make known to hostile elements just what part it could play in military operations."[34]

Patrick's role in helping the Air Service "find itself" is a defining moment in the history of American air power. Prior studies ignore Patrick's contribution to the doctrinal shift in the Air Service in the early years of his tenure. Historians often cite 1926 to 1931 as the "critical turning point" in the renaissance of air doctrine.[35] There is no denying that this period saw a significant burst of intellectual effort on the doctrinal front, but it was Patrick who in late 1921 initiated his campaign to employ doctrine as justification for his vision of a more autonomous and powerful Air Service.

Patrick's credibility with the War Department, General Staff, and Congress set the stage for this eventual understanding and acceptance of a new Air Service doctrine. Patrick knew that doctrine was the point of departure that justified all air arm activities. As such, as part of a 1 December 1921 Air Service Headquarters reorganization, Patrick directed that the newly established Training and War Plans Division, headed by Lt. Col. James E. Fechet, produce a new Training Regulation concerned solely with doctrine.[36] This reorganization was also significant in another sense: it was from the old Training and Operations Branch that Mitchell had wielded his power against Menoher. Patrick, in effect, split the two functions.

Patrick's reorganization memorandum specifically tasked the Training and War Plans Division with "formulating tactical methods to be employed and recommending tactical doctrines for air use."[37] At this point, it would have been precipitous of Patrick to direct the formulation and compilation of anything other than the General Staff–sanctioned tactical mission. What is significant is that the formulation of doctrine was specifically included as a headquarters task.

Patrick's initiative was helped along on several fronts. He was unwittingly assisted in this endeavor by the War Department and directly aided by Maj. William C. Sherman, an instructor at the Air Service Field Officers School at Langley Field, as well as by Billy Mitchell.[38] The War Department's assistance came in the form of an October 1920 order requiring that all individual service training publications be issued as a new Training Regulation (TR) series. The new Air Service Headquarters Training and War Plans Division assigned the TR requirement to the Air Service Officers School where Major Sherman, in turn, wrote what would ultimately be entitled *Fundamental Doctrine of the Air Service*.[39] The initial manuscript was sent to headquarters in May 1921. Sherman's vision of two distinct classifications of military air power, "air service" and "air force," would be the inspiration for the part of Patrick's 1921 report where he unabashedly stated that 80 percent of the Air Service should be dedicated to an offensive mission.[40] Such a notion was anathema to General Menoher, the previous Air Service chief, who had received the same recommendation from Billy Mitchell in April 1919.[41] By contrast, Patrick was fully in favor of this doctrinal concept, and his verbatim use of the idea in his 1921 Air Service report to Secretary Weeks was the first official endorsement of this offensively oriented doctrine at that level.[42]

Patrick was setting the stage to more fully enhance and justify the unique capabilities of the Air Service. This, in turn, would be followed by more strident calls justifying a more independent course for the air arm. Patrick underpinned his campaign with a new doctrinal message emphasizing the offensive nature of the Air Service, but he was careful to keep his message strictly within official channels. The official channels that he used to such good effect, besides his annual reports, were the twice-yearly lectures he personally presented to the Army War College classes at Washington Barracks, his testimony before Congress, and his many official trips out to the field.[43]

Patrick's Army War College lectures are a prime example of his belief in a more dynamic and offensive-leaning doctrine. In his second address to this forum, in April 1922, Patrick emphatically declared, "Briefly and in simplest terms, the mission of the Air Service in war is to gain and maintain aerial supremacy."[44] During the same presentation, Patrick went on to criticize the War Department for the "ludicrously inadequate" attention given the bombing mission, "there being but four squadrons when, at a minimum, there should be twenty-four with another twenty-four pursuit squadrons to clear the way of enemy planes in order that the bombardment planes may work, but to also act as a part of the fighting force." Patrick's mission for the bomber force was the destruction of military targets "both in the theatre of operations and in the enemy's zone of interior." Patrick was also quite clear, even at this early date in his tenure, that "none but observation units should be assigned to ground troops. All other units should be under the Commander in Chief for concentration as he may wish." Patrick summed up this rather forceful presentation to his audience of Army colonels and lieutenant colonels by stating, "Air Service tactics are essentially offensive. We cannot wait to be attacked. Our own air force must strike at that of the enemy, must strike quickly, and must strike hard." Patrick was the first Air Service chief to make the offensive role the primary mission of the service.[45]

Patrick's presentation to the Army War College class was true to his vision of an air arm that was divided into "air service" and "air force." The "air service" of an army would be strictly observation units, which would carry out photographic and reconnaissance functions. The "air force," made up of pursuit, bombardment, and attack aviation, would be purely offensive.[46] Another significant theme was that of "concentration." Not only was Patrick the first head of the Air Service to support the concept of emphasizing an offensive role for the Air Service, he was also the first chief to emphasize the centralized command of the offensive "air forces" under a commander in chief, something not actually realized until the General Headquarters (GHQ) Air Force concept was put forward in 1933 and finally implemented in 1935.[47] Patrick was not yet at the point where, even within military channels, he could advocate outright autonomy, but he was purposefully and methodically laying the groundwork for the ultimate implementation of his new vision.[48]

Patrick's offensive vision was prompted by a renewed appreciation of what could be accomplished by military aircraft. It is obvious that he changed his mind regarding the role of the airplane and the Air Service between the end of the war and the time he again took over the Air Service in 1921. What is not so clear is what prompted this change, but given his former association with the Air Service (regardless of his pronouncements that he wanted nothing more to do with the Air Service), Patrick would have maintained more than a passing interest and was doubtless impressed with the latest Air Service capabilities. Regardless, he needed all the justification he could muster to maintain the hard-fought combat arm status of the Air Service brought about by Congress with the Army Reorganization Act of 1920.

Doctrine, in part, justified autonomy. Patrick was witness not only to the offensive capabilities of the Air Service during the war, where he endorsed the inter-allied independent strategic bombing mission, but he was also impressed with the results of the various bombing trials conducted by Billy Mitchell.[49] The results of these bombing trials gave Patrick substantive justification for his assault on changes in doctrine and his next target: increased personnel and funding.

The second individual (after Mitchell) who Patrick wanted to see when he took over as Air Service chief was Capt. Roger Volandt, the budget officer.[50] Volandt's briefing and Patrick's tour of Air Service units and supply depots convinced Patrick that the Air Service was, indeed, in desperate financial straits and not alone in this regard.

Patrick took over the Air Service at a time when almost every facet of American life prohibited any increase in military spending. Fiscal economy played a large role here but there was a significant pacifist movement within the populace as well.[51] The post–Great War society yearned for "normalcy" and the fiscally conservative Republican administrations of Harding and Coolidge obliged that desire by supporting a token military establishment.[52] Harding, in office when Patrick took over as Air Service chief, was as parsimonious with his words as he was with taxpayers' money. In his message to the first session of the 67th Congress on 12 April 1921 Harding stated, "The Army Air Service should be continued as a coordinate, combatant branch of the Army and its existing organization." For added emphasis he later remarked, "[A]viation is inseparable from either the Army or the Navy."[53] Patrick had his work cut out for him, for he not only faced the daunting challenges of President Harding's fiscal austerity, he also had to contend with the results of the Budget and Accounting Act of 1921, the executive branch enforcer of economy in government. This act created the Bureau of the Budget, which oversaw a unified executive branch budget. The ultimate result was that the Budget Bureau had approval authority over War Department appropriation requests before those requests were sent to Congress. The Budget Bureau, in effect, had the final say over the executive branch budget, for the law required that all executive departments accept the bureau's judgment. It was "the vehicle of the chief executive's [fiscal] policy."[54] Patrick had no recourse to appeal, for when the Budget Bureau spoke the War Department had no choice but to listen, and any appeal or recourse to Congress was therefore nullified.[55]

The assets of the Air Service were also easy pickings for a Congress that blithely ignored the acts it passed into law. For example, even though Congress enacted the Air Service Act of 1920, which authorized the training of 2,500 cadets per year, scarcely 200 were schooled within the year following the act's passage. Officer allocations were fair game as well; 1,516 were authorized, and by 1923 only 883 were on duty. Thus Patrick was well aware of the dangers in dealing with a whimsical Congress that would authorize funds on one hand but would not appropriate them on the other. Patrick knew that getting more funding to make his Air Service viable would be at the expense of others, meaning the War Department and the Department of the Navy.

Patrick and Mitchell were of one mind in this particular regard, but initially at opposite poles when it came to how much the Air Service could contribute to America's defense. Whereas Mitchell spoke of making the battleship obsolete and displacing the infantry as the arbiter of battle, Patrick stated that the mission of the Air Service in war was "to assist the other branches of the service in bringing the war to a victorious conclusion."[56] Major Sherman thought much the same way, emphasizing that once one controlled the air, the mission of the air force was to contribute to the destruction of "the most important enemy forces on the surface of the land or sea."[57] The leadership of the Air Service was, at this early date, clearly interested in the application of lessons learned from the Great War. The formulation of those lessons into doctrine would, in turn, support Patrick's requests to obtain increased funding for new technology in the form of state-of-the-art aircraft for the offensive mission. Patrick was so committed to the Air Services' offensive mission that in January 1922 he returned $1,399,000 in aircraft procurement funds to Congress because the planes to be purchased with this money would be wholly unsuited for offensive operations. But he attached some strings to this offer with the provision that the funds would revert to the Air Service budget in 1925 when a new generation of aircraft would be available.[58]

Patrick's blueprint for saving the Air Service, and then enhancing its mission, was well thought out. In the first full year that he was in charge, he concentrated on controlling Mitchell and instilling discipline within the Air Service, coupled with his activist agenda regarding doctrine and budget enhancement. As far as doctrine was concerned, Patrick was the first Air Service Chief to have formalized (in War Department–approved TR 440-15) the concept that an "air force" had a role coequal with ground forces.[59] With regard to the Air Services budget, he consolidated supply depots and even returned part of the acquisition budget to Congress. He won over his own personnel as well. He was by no means a grim taskmaster. His administrative style showed compassion and resourcefulness.

From the first day of his tenure, he continually fought for the interests of his civilian as well as military personnel. When Patrick's budget officer asked about deducting the time for civilian employees who could not get into work following a blizzard, Patrick told him to pay them in full and followed it up with instructions that employees who did make it to work could leave for home early.[60]

Having increased the sympathy of the General Staff and Congress, Patrick had room to maneuver when he presented his annual report for 1922 to Secretary of War John Weeks. Patrick and Secretary Weeks had a very cordial relationship, made even more congenial when Patrick saved Weeks from an acutely embarrassing episode. When one of the very infrequent aircraft manufacturing contracts was awarded in the summer of 1922, Weeks had his picture taken presenting the signed contract to the congressman who represented the district where the factory was located as the company president looked on. The following week the successful bidder had an office call with Patrick and proudly exhibited the picture and related that the congressman planned to use it in his reelection campaign to bolster his support among his working-man constituency. Patrick immediately saw the dire implications: the picture could be construed to mean that political pressure had been brought to bear in the award of the contract. Patrick collected all copies to include the original photographic plate of the picture and presented them to Weeks with the explanation that this would undermine Patrick's effort to run the Air Service "along proper lines."[61] What Weeks had originally considered a favor to a long-time personal friend (the congressman) took on new meaning with Patrick's keen awareness of the political scandal that could be made of the incident. The pictures and the plate from which they were made were destroyed. Weeks never forgot Patrick's well-reasoned and thoughtful gesture.

Weeks was also thankful to Patrick for keeping Billy Mitchell under wraps. From Patrick's first day as chief up through 1922 and well into 1923, Mitchell's intemperate outbursts ceased. Patrick reviewed every Mitchell speech and article as required by the understanding brokered under the watchful eye of General Harbord. When Mitchell traveled on official business, Patrick required that Mitchell report back to him on a regular basis and to say nothing that could be misconstrued as criticism of the War Department.[62] All in all, Patrick's management of the Air Service during his first year created a bond of mutual trust between himself and Secretary Weeks. Thus, when Weeks received Patrick's 1922 annual report, he gave it more than due consideration.

Patrick's 1922 Annual Report and the Lassiter Board

In Patrick's mind, there was never a doubt regarding the efficacy of air power as demonstrated by the famous bombing trials off the Virginia Capes in the summer of 1921 when Billy Mitchell and his 1st Air Brigade sank the *Ostfriesland*.[63] Patrick knew that the Joint Board report on the bombing trials, even though signed by his good friend General Pershing, was a document of compromise. Just as Pershing was a man of compromise, so was Patrick: "he advocated air power with moderation, reason rather than emotion."[64] Given that arguing from reason was one of Patrick's strong suits, it is surprising that he won any early concessions at all to improve the Air Service. President Coolidge spoke for many in the nation when he remarked: "Who's gonna fight us?"[65] Coolidge directed his War Department officials to find ways to reduce costs "without weakening our defense but rather perfecting it."[66] This simple idea of "perfecting" the nation's defense early on formed the basis for Patrick's argument for an improved Air Service. Patrick, speaking in January 1923, employed this concept long before Coolidge did, suggesting that a more efficient defense could consist of the decisive use of air power at the start of a conflict.[67]

Unfortunately for Patrick, the Air Service had been hamstrung by huge stocks of obsolete World War I materiel. Except for the purchase of a few prototype aircraft to meet specific requirements, the Air Service could in no way meet wartime requirements with the aircraft they had on hand. Directed by Congress to use war surplus DH-4s powered by Liberty engines until the stocks were exhausted, it committed the Air Service to the operation of obsolete aircraft well into the late twenties.[68] This requirement grievously hampered the growth of the commercial aircraft establishment. In Patrick's mind a modern Air Service with sufficient planes, personnel, and armaments was the ideal "economy" measure, and he made his 1922 annual report the cornerstone of his long road map to legislation that would ultimately cure his service's long-neglected ills.

Patrick expounded on two major themes in the introduction to his 1922 annual report: the importance of a viable combat aviation force and the requirement for a strong commercial aircraft establishment to support that force once war was declared.

The 80 percent [of the Air Service] devoted to "air force" or "combat aviation" has suffered. The need for increased strength in this vitally important arm is readily apparent and "urgently recommended.

The growth and development of civil aviation constitutes an invaluable adjunct to military preparation in time of peace. The stimulation of competition resulting from an extensive use of aircraft for commercial purposes will ensure the greatest progress in design and production and will . . . provide a manufacturing industry capable of meeting the increased demands of national defense in case of an emergency.[69]

Patrick made his desires plain but, more important, he made them in a temperate manner. His approach to the problems of the Air Service can best be characterized by a representative quote from his 1924 congressional testimony:

There are, on the one hand, enthusiasts who believe that the coming into being of aircraft have practically scrapped all other combat agencies; and on the other hand, conservatives who consider aircraft as merely auxiliaries to previously existing combat branches. The truth, of course, lies somewhere between those two views.[70]

Patrick's approach was much appreciated by the congressmen he addressed, as one committee representative noted when he thanked Patrick for favoring "evolutionary rather than revolutionary" changes in the national military organization.[71]

Given Patrick's assertive speeches, articles, and testimony in 1921 and 1922, culminating in his blunt annual report of 1922, few students of this era have fully realized the degree to which Mason Patrick enthusiastically supported the more radical ideas of his service arm. There is little appreciation of how he maneuvered, in his individual style, to minimize the animosity that the General Staff and War Department had for the Air Service, and, in doing so, brought a measure of credibility to his service. These early actions did much in the coming years to justify an increase in missions and a corresponding growth in the Air Service budget.

At the end of Patrick's first year, his annual report was a blueprint for the future of the Air Service. The testament to his first year can best be summed up by the key issues addressed in that report. It pointedly emphasized the offensive nature of the Air Service, the development of commercial aviation plus the manufacturing facilities for same, and his service's commissioned personnel shortage. In addition, Patrick's numerous administrative "economy" initiatives and his effortless control of Billy Mitchell demonstrated his competency as a manager and a team player.

Overall, what Patrick accomplished in 1922 was the genesis for change regarding how the War Department and the General Staff viewed the future role of the Air Service. In particular, his 1922 annual report was the genesis for what ultimately became the 1926 Air Corps Act. This singular document, strongly endorsed by Patrick, more than any other factor influenced the War Department to seriously explore the deficiencies of the Air Service. Patrick was directly responsible for calling attention to the need for appropriate corrective action, which ultimately resulted in a significant long-term consequence for the Air Service: the path to independence.

7. The Lassiter Board and "Fundamental Conceptions": Spadework for the 1926 Air Corps Act

A Page from Mitchell

Patrick's actions in 1922 planted the seed for what would ultimately become the 1926 Air Corps Act; over the next two years he would work to ensure that his plans had continued viability. At the end of 1922, Patrick left no doubt with his superiors that he was entirely dissatisfied with the condition of the Air Service. The major points of his report to the secretary of war were condensed in numerous articles and speeches wherein Patrick highlighted his service's severe shortcomings.[1] Most important, these speeches were not confined to professional military audiences. Civilians from Philadelphia, Akron, Chicago, and Los Angeles heard his well-reasoned arguments in 1923.[2] In all, Patrick spoke in front of a civilian audience on the average of once a week in 1923, but this satisfied only a fraction of the invitations he received requesting a talk or lecture on airpower issues of the day.[3] Patrick's correspondence file overflowed with a familiar notation in his blue pencil instructing his executive officer: "Send regrets, MMP."

Patrick was busy on the lecture circuit, as well as a regular contributor to the national press. The *New York Times,* by far, received the majority of his articles, followed by Washington area newspapers. Sometimes a newspaper requested an article addressing a certain issue; occasionally Patrick would forward an unsolicited piece. When Patrick addressed the general public, either in person or in the press, his approach emphasized the great importance of aeronautics (both military and civil) to the future of the nation.[4]

While Patrick cultivated both the civilian press and the American public, he did not ignore the military or the American defense establishment. Quite the contrary: on 15 January 1923 Patrick delivered the keynote address at the American Defense Society in New York City.

In the next war if one side secures supremacy of the air, that side will almost inevitably be the victor. . . . It follows that our national safety demands that we prepare accordingly . . . [with] what may be called the air force, consisting of pursuit, attack, and bombardment planes . . . called upon to execute missions entirely independently of any ground forces. With such an air force ready to strike on "D" day, the day when war is declared, and able to cope victoriously with the enemys air force, there is the possibility that the ground troops might never have to fire a shot.[5]

It was clear that Patrick now saw the airplane as a revolutionary weapon, capable of much more than its primary World War I role of reconnaissance. If aircraft were to be used properly, their employment should take advantage of their inherent offensive capability under centralized command.[6]

Support from the Rank and File

Billy Mitchell could have been the author of the speeches that Patrick was making in 1922 and 1923, and in particular the American Defense Society speech, but he was not even in Washington at the time.[7] Patrick had been highly successful in keeping Mitchell preoccupied and out of Air Service headquarters and Washington affairs. Mitchell was so deeply involved with numerous trips to the field to visit Air Service bases and the technical center at McCook Field in Dayton, Ohio, that he spent very little time in Washington. In fact, after returning to the states in March 1922 from his four-month European tour, Mitchell was not listed as having attended any regular Monday morning staff meetings until 30 July 1923.[8] On the occasions that Patrick was away from the city and Mitchell was the senior officer at headquarters, Patrick kept close tabs. His faithful executive officer from October 1921 to June 1923 was Maj. William H. Frank, who kept Patrick fully informed of office business with letters and telegrams.[9]

Patrick had other reasons to keep a close eye on Billy Mitchell. Mitchell had personal problems, which, to Patrick's benefit, also contributed to Mitchell's absence from Washington. Mitchell's first marriage ended in an acrimonious divorce in 1922. As early as December 1921, Patrick became personally involved in this domestic quarrel when, just prior to Mitchell's departure for the four-month European tour, Caroline Stoddard Mitchell accused her husband of being mentally unstable. Unfortunately, Mrs. Mitchell chose to address her accusations to acquaintances on the General Staff, in particular General Harbord. Patrick's immediate intervention permitted a quick resolution, and Mitchell soon departed on his journey.[10] In addition to his marital misfortunes, Mitchell lost his mother, with whom he was very close, in December 1922.

When Mitchell was in the States, Patrick kept him busy on mostly technical research and development matters, an issue that greatly concerned both individuals, as well as racing events and publicity-rich public affairs work. Patrick was personally involved with the Air Service's public affairs program and never missed attendance at the many yearly aircraft races, including the prized Pulitzer races.[11]

Patrick and Mitchell worked well together through 1922 and 1923 and did so with the help of a competent and enthusiastic headquarters staff led by Lt. Col. James E. Fechet, a loyal and no-nonsense cavalryman who had learned to fly in 1917. In 1921 Fechet filled Mitchell's job as chief of the Training and War Plans Division, which Mitchell had held under Menoher. Fechet was a friend of Mitchell's and would later replace him as assistant chief of the Air Service when Mitchell's appointment expired in 1925 before going on to become chief of the Air Corps with Patrick's retirement in 1927. The remainder of Patrick's staff were all Air Service AEF veterans who had fairly common operational flying backgrounds and knew each other well.

Patrick welcomed his staff's opinions on all aspects of Air Service operation. The minutes (although at times incomplete) of Patrick's weekly staff meetings reveal an atmosphere that welcomed freewheeling debate on all aspects of an issue, at the end of which Patrick would almost invariably resolve the issue with appropriate orders issued to the responsible parties.[12]

One topic of discussion during 1922 revolved around Patrick's annual report, which was thoroughly staffed through each division. Consensus was a strong point with Patrick. Like Pershing, he wanted team players, but by no means did Patrick want "yes" men. And given Patrick's earlier reluctance to appreciate many "radical" ideas within the Air Service, Patrick expected anything *but* "yes" men. Patrick's doctrinal metamorphosis in 1922 was a pleasant surprise to the tight-knit band of fliers, one of whom was Capt. Ira Eaker. Arriving at Air Service Headquarters in June 1924, this future three-star general and World War II Eighth Air Force commander served as Patrick's assistant executive officer and wrote many of his speeches. In later years Eaker always spoke highly of Patrick, even citing Patrick as responsible for the Air Corps being in a much better position at the start of World War II than would otherwise have been the case.[13]

Patrick also gave credit where it was due. In a January 1922 speech he stated that "all credit" should go to General Mitchell for the "remarkable" results of the previous summer's bombing tests, which, in Patrick's words "demonstrated beyond a doubt that aircraft were capable of . . . destroying any surface ship which has yet been designed."[14] As Patrick vigorously supported his officers and men, they in turn returned the favor, both at Air Service Headquarters and out in the field.

At the start of 1923 the officers and men of the much maligned and abused Air Service were thankful for the change in leadership at the top. Now they had a chief who believed as they believed. They had a chief who spoke clearly and got visible results. "Patrick was an intelligent, sympathetic leader who understood the needs of the Air Service and knew how to convey them to the General Staff. . . . [H]e shared [Mitchell's] enthusiasm for developing the air arms roles, missions, and capabilities. Veteran Army aviators respected him."[15]

Respect for Patrick was not limited to the Air Service. His views on the Air Service and aeronautics in general were solicited by Washington's key politicians and bureaucrats.[16] Patrick was on particularly good terms with Assistant Secretary of War J. Mayhew Wainwright. An aviation enthusiast, Wainwright wrote several articles dealing with aviation issues, and agreed wholeheartedly with Patrick that a national aviation policy/organization was necessary. In addition, by invitation, Patrick regularly lunched with senior members of cabinet departments.[17] He used these meetings to seek understanding and support for his impending fiscal and ideological battle to save his Air Service. During the summer and autumn of 1922 he grew more vocal in his attempts to bring attention to the Air Service's death spiral.[18] At this point "Patrick felt the Air Service had been practically demobilized and could no longer meet peacetime demands, much less any national emergency."[19] Patrick voiced these concerns in the 1922 annual report.

The Secretary of War Responds

On 18 December 1922 Secretary of War Weeks directed Patrick to present a detailed study that would be the basis for remedial action to fix the deficiencies noted in his report.[20] Prior to presenting what would ultimately become his testimony before the Lassiter Board, Patrick was extremely reluctant to oblige unless the War Department accepted Patrick's premise that Army aviation be "properly balanced" into an "Air Service" ground-support force at 20 percent and an offensive "Air Force" component of 80 percent. Otherwise, his eighteen-month war organization study, then in progress, would have to be scrapped.[21] Whereas the War Department wanted to work from the basis of a new Air Service "peace establishment" to meet "the approved war time expansion" requirements, Patrick suggested just the opposite.[22] Patrick pointedly responded that since he believed that both the peace and wartime plans for the Air Service were "incorrect," his proposed war plan for the Air Service (which he attached) was the only appropriate starting point.[23] Prior to his Lassiter Board testimony, Patrick made his case before the Air Service and aviation community at large by publishing excerpts from his annual report in *U.S. Air Service*.[24]

As if his ultimatum to the War Department were not enough, Patrick added more organizational and doctrinal fuel to the fire. He wanted not only to change numbers and type of aircraft but also to modify the way in which they would be used. He was particularly sensitive to the centralized command of forces that would remain in a strictly ground-support role. The observation units that would remain with ground commanders should be removed from divisional command and placed at corps level. Even more important is Patrick's insistence that "[v]ery often, there is as distinct and definite a mission for the Air Force independent of the ground troops as there is for the Army and Navy independent of each other."[25] Doubtless based on his involvement with the combined Allied strategic bombing effort during World War I, Patrick also noted, "Bombardment aviation especially will act with ground troops only in very rare instances."[26]

Patrick was attempting to shift the Air Service to a centrally controlled offensive "Air Force" mission with two major changes: (1) the withdrawal of all Air Service assets from division level with a commensurate consolidation at corps; and (2) the provision for an adequate, well-balanced Air Force concentrated in a General Headquarters (GHQ) Reserve, to include a six-fold increase in bombardment groups. The rationale he offered was based partly on economy of forces but also on the offensive promise of an Air Force.[27]

Patrick's 7 February 1923 response to Secretary Weeks concluded, "A detailed discussion of the proposed organization would make this a voluminous endorsement. The importance of this matter warrants a conference at which I shall be glad to explain the numerous details."[28] Patrick's response to Secretary Weeks resulted in the Lassiter Board, a group of General Staff officers headed by Maj. Gen. William Lassiter, to hear Patrick out.

The Lassiter Board: Basis for the Air Corps Act of 1926

The Lassiter Board convened early on Thursday morning, 22 March 1923, and was to meet in daily session until its job was finished. In all, it took only six days. On the face of it, it would seem that Patrick's proposal had no chance of surviving the scrutiny of a board composed of seven General Staff officers.[29] But it was also composed of officers whom Patrick knew, and some he had worked with during AEF days. General Lassiter (U.S. Military Academy Class of 1886) was the G-3 (Chief of Operations and Training) and had known Patrick when they were cadets at West Point. Lassiter had also been a member of the April 1919 Dickman Board, the results of which tied aviation units closely to the ground effort.[30] In addition to Lassiter, there was Brig. Gen. Briant H. Wells, the chief of the War Plans Division (a division directly subordinate to G-3); Brig. Gen. Stuart Heintzelman, the G-2 (Intelligence); a colonel from the Quartermaster Corps; and a lieutenant colonel from coast artillery detailed to the General Staff.[31] The only members who had any association with the Air Service were Lt. Col. Frank P. Lahm, a conservative airman (who had butted heads with Billy Mitchell in France) who was assigned to G-3 on the General Staff, and Maj. Herbert A. Dargue, an outstanding aviator who was on duty with Patrick's staff, as the committee's recorder.[32] Dargue and Lahm were two of the earliest pioneer aviators still on active duty. Rounding out the membership on the board was Brig. Gen. Hugh A. Drum, who, when he was Chief of Staff, First Army, AEF, was sorely at odds with Billy Mitchell and his "ill-disciplined" flyers. Drum wanted nothing to do with anything that smacked of Air Service independence. In fact, when the August 1919 Menoher Board was considering the impact of the New and Curry bills and solicited comment from senior AEF officers, General Drum was unequivocally critical of the possibility of Air Service independence and of Air Service officer discipline.[33]

General Lassiter set the tone for the board as well as the agenda when he proposed that the board "get the picture that General Patrick has in his mind and then go back and take it up in detail." From that point on, until the board's conclusion, Mason Patrick was the center of attention. Patrick began by explaining the efficiencies that would result by reducing the support (mainly observation) assets of the service and increasing the offensive assets, especially pursuit and bombardment. This was followed up with a discourse on concentration of force and its centralized control: "The principle of concentration of air forces becomes a maxim . . . and I am convinced that the concentration of all air force under one GHQ Reserve Commander is the most effective way of assuring aerial supremacy. . . . To assign attack or bombardment to the army permanently is a waste of material."[34]

While all board members were willing to give Patrick a fair hearing on his proposal to attach air assets to the corps, as opposed to the division, Drum opposed change from the start. He was adamant that the board apply the guidance employed by the General Staff War Plans Division to properly determine wartime organization.[35] "All aviation in an Army should be employed for participation in the battle, and all strategical bombardment and reconnaissance should be done by aviation in the G.H.Q. Reserve."[36] Drum also saw no value in bombardment. He emphatically stated, "I see three functions of the Air Service: coast defense, air defense, and operation with land forces."[37] In the end, the vote was four to three to keep the observation squadrons at division, corps, and army level.[38]

With regard to Patrick's other issues, the board was more understanding. It agreed that an air force comprised of attack and pursuit squadrons should be an integral part of each field army, and an air force of bombardment and pursuit assets should be assigned to a GHQ Reserve for the missions that Patrick deemed "independent."[39] This was a crucial victory for Patrick: the Army formally acknowledged the concept of independent action by air force assets. It was a necessary prerequisite for Patrick's much more important quest to have the Air Service augmented to fully support the Six Army Plan then in existence. This contingency war plan projected the fielding of six armies within six months of "D-day." Patrick argued persuasively that (1) based on World War I experience, it would take much longer to field a commensurate Air Service organization to support same; and (2) if the full complement of Air Service assets existed at the outbreak of war, they would be immediately effective, and perhaps decisive, against a potential enemy attack.

To be effective in such a scenario, Patrick presented an organization plan worked up by his staff. It called for pursuit groups to be increased from fourteen to twenty-four; bombardment increased from one to six; ground-attack squadrons reduced from seven to six; and observation squadrons reduced from 118 squadrons to 100. The proposed increase in aircraft numbers was necessary for two other reasons: the American aircraft industry was on the verge of vanishing and "the larger part of the aircraft now on hand is war produced and is deteriorating rapidly."[40] Patrick wanted to turn this situation around and was steadfast when testifying about the restructuring, especially about the ideal premise that observation assets based at corps and army level would offer greater flexibility against enemy threats. To increase the Air Service mission even further, Patrick recommended that the coastal defense issue also be reconsidered.[41] He suggested that coastal defense be carried on by the Air Service out to 200 miles, with the Navy limited to open ocean duties. Patrick's point, again, was one based on economy of forces: "It provides that where the Navy's interests are paramount the Navy should control; when the interests of the Army are paramount the Army should control. In connection with the coast defense there is, I think, a decided possibility of duplication. . . . [It] will be one of the greatest gains to efficient national defense."[42]

Patrick's resolute manner and his presentation to the board were doubtless impressive. The committee report that was signed by Secretary Weeks said in part:

The Committee finds our Air Service to be in a very unfortunate and critical situation. . . . For lack of business our aircraft industry is languishing and may disappear. . . . We cannot improvise an Air Service and yet it is indispensable to be strong in the air at the very outset of a war.[43]

Patrick did not get everything he desired, but the things he did get were critically important stepping stones to independence. The board retained keeping observation units at division, corps, and army level, but they also complied with Patrick's plan in several important ways: (1) an observation reserve and an air force of attack and pursuit aircraft should be dedicated to each field army, with a commensurate reserve under General Headquarters; and (2) an air force of bombardment, pursuit and airships should be directly under General Headquarters for special and strategical missions with, or independent of, ground troops and operating in large units to ensure great mobility and independence of action.[44] The board agreed with Patrick's proposal for aircraft augmentation with the following configuration: instead of the one bombardment group at GHQ Reserve, there should be two bombardment groups and four pursuit groups. Last, and perhaps second in importance to Patrick (behind the independent mission issue), was the recommendation that Congress make annual appropriations of $25 million for the Air Service each year for the next ten years and that $15 million be earmarked for the purchase of new aircraft. This amount was exclusive of any "pay, upkeep, and housing of personnel."[45]

On 27 March 1923, after six days of testimony (almost all of it by General Patrick) and a review of pertinent documents, the Lassiter Committee submitted its report to Secretary of War Weeks. But Weeks did not rubberstamp it. He had some questions, which he put to Patrick on 4 April.[46] Patrick responded the next day with a lengthy memo that was unusual for its unique format. In place of what would normally be a title, the memo had, in capital letters and boldface type, the following, centered at the top of the page: ARMY AIR SERVICE PRACTICALLY INOPERATIVE. COMPELLED TO USE WAR TIME AIRPLANES WHICH WILL SOON BE EXHAUSTED. EFFICIENCY OF AVIATION CORPS HAMPERED BY LOW RANK OF ITS OFFICERS. INEQUALITIES OF PRESENT ARMY PROMOTION LIST WORK INJUSTICES TO FLYING OFFICERS.[47] It was not Patrick's choice that his testimony before the Lassiter Board primarily concerned organizational issues and the need for new aircraft. The board members framed the issues. They did not discuss the promotion and rank problem of the Air Service, even though Patrick had mentioned it in his annual report. On the Army's Single Promotion List published in the winter of 1921 the majority of Air Service officers, especially those below the rank of major, were at the bottom. For the most part, this was due to the extended training time required of Air Service officers until they were commissioned. These officers were outranked by their peers in other branches who had even less active service time.

To compound the problem, when fatalities occurred within the Air Service, especially in field grade ranks and captain, which was not infrequent, the promotion billet that thus became available benefited all other branches first, on a seniority basis. As Patrick pointed out, "The obvious shortage in the field grades in the Air Service can never be remedied as long as the present Promotion List is effective, since many of the first lieutenants can never hope to reach field grade officer prior to retirement or death from aircraft accidents."[48] To remedy this problem, Patrick favored a separate promotion list for the Air Service, much as the Army Medical Corps had in place, but the General Staff was opposed to this for reasons of economy, overall morale, and ease of management.

Both Wainwright and Weeks were eager to bring some respite to the Air Service and were in agreement with the Lassiter Board report. But the War Department was willing to go only so far. Secretary Weeks was in agreement with the first two items listed in Patrick's 5 April memorandum concerning the overall abysmal state of the Air Service and the obsolescent condition of the aircraft. But as far as the problems stemming from rank and promotion were concerned, both Weeks and Wainwright were silent. Weeks approved the Lassiter Report as written; no mention was made of the promotion issue.

Results and Consequences of the Lassiter Board

Two immediate effects of the Lassiter Board concerned how the Air Service would be viewed as a combat arm within the overall organization of the Army and the importance of pursuit aviation within the Air Service itself. This new understanding was codified in the new Army Field Service Regulations for 1923: "no one arm wins battles."[49] The Field Service Regulations, of course, assigned observation units to division, corps, and armies as previous tradition and the Lassiter Board recommended. But the new Field Service Regulation treated pursuit aviation with something akin to reverence: "[Pursuit aviation] created the conditions which enable other [combat] elements to operate with the greatest degree of effectiveness" and was thus "the most vital element of the air service."[50]

Patrick had won a hard-fought victory. Formal War Department recognition of the offensive capabilities and mission of pursuit and bombardment assets went hand in hand with the increased aircraft augmentation recommended by the Lassiter Board. An authorized increase in new pursuit and bombardment aircraft for the offensive mission was one of Patrick's primary goals, which would also enhance the aviation manufacturing base. Things had gotten so bad that even the rumor of an impending contract to recondition a limited number of war-vintage DH-4 aircraft brought a stream of manufacturing lobbyists to Patrick's door.[51]

Patrick's desire to alleviate the desperate conditions within the aviation manufacturing community was not unknown. In fact, it was a well-publicized part of his agenda. As early as 1919, when Patrick wrote to Pershing with suggestions for Pershing's upcoming congressional testimony regarding the future Army reorganization, Patrick suggested federal regulation of civil air travel and federal assistance to maintain a strong aviation manufacturing base run by dedicated aviation experts. His suggestion was in response to his bitter experience with the unfulfilled exaggerations and gross inefficiency of the automobile manufacturers who turned into wartime aviation builders.[52]

For Mason Patrick and the Air Service, the Lassiter Board resulted in an extremely important incremental success. Patrick was well aware that complete victory (complete centralized command of air assets, not to mention the Single List promotion issue) was impossible at this point.[53] He needed to keep the Air

Service alive and the military aircraft manufacturers viable. With a new "independent" mission under the wings of the Air Service, Patrick's doctrinal epistle was accepted as gospel and signed by the secretary of war.

"Fundamental Conceptions of the Air Service"

Once the Lassiter Board accepted the majority of his recommendations and Secretary Weeks signed the Lassiter Board Report, Patrick had his staff draft a paper entitled, "Fundamental Conceptions of the Air Service: Prepared Under the Direction of Chief of Air Service, 1923."[54]

Patrick was on an inspection tour of Air Service squadrons across the country while his staff drafted the paper.[55] His absence had no impact on the paper's preparation; all elements within "Fundamental Conceptions" had been voiced previously by Patrick. What Patrick wanted the document to accomplish was the formal codification of the new Air Service doctrine as just agreed to by the Lassiter Board and Weeks. He was not going to wait for the General Staff to dictate via new Field Service Regulations what the new doctrine would be. Patrick interpreted the acceptance of his ideas as a clear signal to press forward with their formal implementation.

Patrick did not waste any time worrying about the outcome of the new regulations. Patrick's last day of testimony before the Lassiter Board was 26 March. The board forwarded the finished report to Weeks the next day. Patrick had a meeting with Weeks on 4 April and forwarded a follow-up memorandum to him on the fifth. Patrick then departed Washington for a tour of bases to include Ohio, California, Texas, and Kansas, and did not return to Air Service Headquarters until 10 May.

While Patrick was on the California leg of this trip, he visited Rockwell Field, which was under the command of Maj. Hap Arnold. Patrick was very impressed with Arnold, especially for his innovative work with Forest Fire Patrols and other public relations programs. Patrick was so impressed with the inspection he added a commendation to Arnold's official (201 personnel) file. Just over a year later, Patrick would recall Arnold to Washington to attend the Army Industrial College, followed by duty on the headquarters staff as chief of the Information Section. But when Patrick flew into Rockwell on this particular tour, he experienced an embarrassing incident that Arnold made the most of in later years to demean Patrick. Upon landing, Arnold had his officers and men assembled for inspection. Patrick, who wore a toupee, lost the hairpiece and the hat that was holding it in place to the prop wash of a passing plane. As Arnold described it: "The prop wash caught the campaign hat and wig. They both went skipping across the field. Everybody just stood there and gaped. They didn't know what to do. Old Patrick says, 'Well, don't just stand there, go get the SOB.' Some guy brings this thing back, holding it like he was holding a skunk."[56]

While Patrick was away dealing with inspections and errant toupees, Weeks had approved the Lassiter Report on 24 April, which Patrick had been duly informed about, and upon his arrival in Washington on the tenth, Patrick signed "Fundamental Conceptions of the Air Service," which was used as the basis for the War Department–approved Training Regulation 440-15. Modified to conform to the "traditional military view," it became TR 440-15, published on 26 January 1926. More than a training regulation, it was the first declaration of Air Service independence from ground forces formally accepted by the War Department. Patrick pledged his fealty to Army doctrine and immediately qualified that fealty. Under Section II of "Fundamental Doctrines," Patrick's document quoted Army Training Regulation 10-5: "The Infantry is the basic arm and upon its success depends the success of the Army. All other branches are organized, equipped and trained to assist the Infantry in its needs, functions, and methods of war."[57] The text of the document immediately following stated: "This basic principle is accepted without qualification and none of the paragraphs that follow are to be considered as qualifying or questioning the thought contained therein."

With that statement as insurance against any charges of radical independence by the General Staff, Patrick went on:

In waging war with this fundamental conception as a basis for all operations, the auxiliary units of the Air Service, that is to say, the Observation units, are definitely identified with ground units and those services which they directly serve. However, the combatant branches of the Air Service, that is, the Air "Force" units, in the furtherance of this basic principle must wander far afield, and there are times when the principle can be served only in an indirect way by these units, who are carrying on such operations as they deem vital and under the immediate Commander in Chief of the Forces in the field.

As the [Army and the Navy] have worked out their doctrines of war in their native elements, so must military aviation . . . and it would be both fallacious and fatal to accept and sustain a tactical doctrine for military aviation, which is built upon an assumed similarity with ground units, and with an attempt to treat it solely as an auxiliary service to the ground army.[58]

The document detailed the inherent independence of action that the "Air Force" would have when assigned to a GHQ, since it reported directly to the General Staff. This was Patrick's declaration of independence, a declaration blessed by the War Department, and one that included, for the first time in an official War Department–sanctioned document, TR 440-15, the term "Air Force." In addition, as historian Thomas H. Greer noted, while TR 440-15 overall supported the "traditional military view," the ideology and course materials at the Air Service Tactical School "described a more independent function for air power."[59]

From all indications, the genesis and revision of TR 440-15 were blessed by and nurtured by Patrick and did not have the imprimatur of Billy Mitchell. During February and March 1923 he was busy shuttling between bases from Canada (testing winter flying routes and conditions) to Texas (conducting tests and maneuvers).[60] Mitchell, though, would have agreed with much of what was in "Fundamental Conceptions" except fidelity to the infantry "queen of battle" tenet of Army Regulation 10-5.

In Search of Air Service Allies

With the approval of the Lassiter Report by Secretary Weeks, the Air Service gained increased respectability, a new independent mission, and the hope for significant new funding. But only if Congress agreed. In an attempt to sway Congress to his point of view, Patrick planned a new public relations campaign, and he made Billy Mitchell a part of it.

Patrick cultivated and won numerous allies in his quest to make the Air Service an indispensable part of America's defense. The Lassiter Board results and the commensurate boost the Air Service received from the newly published *Field Service Regulations* set the stage for Patrick to press his case. A most important audience had always been the senior officers of the U.S. Army. Patrick was well aware that he had to win over this conservative element to his cause. Through the remainder of 1923 and into 1924, Patrick approached this task without "any exaggerated claims, and overstatements."[61]

On the civilian side of the War Department, Patrick had powerful allies. Secretary of War Weeks was sympathetic to the Air Service, and Assistant Secretary J. Mayhew Wainwright was undeniably on the side of the Air Service. Wainwright resigned from his War Department position at the end of March 1923 (to be replaced by Dwight F. Davis) to assume a seat in the House representing the state of New York. One of his parting shots of support for the Air Service appeared in the April 1923 edition of *U.S. Air Service*. In the article Wainwright criticized the inadequacy of the $25 million recommended by the Lassiter Board. Wainwright thought this amount represented the very minimum required just to meet the most pressing needs.[62] Wainwright's article could have been written by Patrick's staff as, point by point, it endorsed each major issue in Patrick's Lassiter Board testimony. Wainwright went on to be a strong advocate for Air Service interests in the House and contributed regularly to popular and professional journals regarding the Army Air Service and aviation in general.

While Patrick did not shy away from publicizing his views, he did so in a cautious manner, preferring that, when practical, others do so, especially if these individuals were respected members of government or business. During and immediately following his annual report and the events that finally culminated in the

Lassiter Board, Patrick was careful not to draw additional attention to contentious Air Service issues by baiting the waters.[63] He did not, however, let Air Service involvement in air races and public relations events decline. One public (and Air Service) relations event that Patrick finessed in the summer of 1923 was a huge boon to Air service morale, and to public relations in general. Patrick learned to fly.

8. Patrick Takes on a Pair of Wings, the Navy, and the Army General Service Schools

Patrick Takes the Stick

Patrick began his flying lessons at age fifty-eight and won his Junior Aviator wings in the summer of 1923 at age fifty-nine, not an inconsequential feat considering the challenging state of the aircraft of the day.[1] In describing the accomplishment, Patrick's pilot and flight instructor, Maj. Herbert "Bert" Dargue, noted that "there is probably no one thing the Chief of Air Service could have done to raise the morale of his officers and men than to learn to fly himself."[2] Patrick received dozens of telegrams, letters, and notes congratulating him for this aeronautical accomplishment.[3] Chief among them were notes from several members of the General Staff (including Pershing), many old friends who were commanding in the field, West Point classmates, key leaders in the civilian aeronautical community, and members of Congress. Adm. William A. Moffett, chief of Naval Aviation, even saw fit to pass on his regards. Conspicuous by its absence was a congratulatory note from Billy Mitchell.

Having qualified as an aviator, Patrick took every opportunity to fly, though always accompanied by another pilot, when he went on his numerous inspection tours. He put a lot of stock in the effect his ability to fly would have on his leadership:

Nothing did more than this continual flying to win the confidence of the men, much younger, with whom I was in contact, trying to direct and guide them in an effort to make of the Air Service a united body of men all working toward one end. . . . To learn first hand their own ideas about planes and motors, the trouble they experienced . . . with the planes provided . . . and their suggestions for the betterment of the Air Service.[4]

The betterment of the air arms mission was a central tenet with this dynamic Air Service chief. When Patrick flew out of Bolling Field on inspection trips and speaking engagements, he usually headed out over the Potomac River and the Chesapeake Bay; in the process of flying over countless ships, he remembered another project on his agenda. Patrick was hard at work behind the scenes working with the Navy and the War Department to accomplish another public relations success. He wanted the Air Service to bomb additional ships. He wanted an encore to the highly successful *Ostfriesland* bombing of July 1921.

In Search of Ships

Patrick quietly negotiated for many months with the Navy for the release of two ships, the *Dakota* and the *Delaware,* to be used for bombing trials.[5] These ships were scheduled to be scrapped due to the requirements contained in the Five Power Naval Treaty as part of the Washington Conference for the Limitation of Naval Armaments.

Both Patrick and Mitchell discussed the issue of acquiring additional obsolete vessels not long after Patrick came in as chief.[6] But Patrick was not anxious to alienate the Navy since he had already challenged them over the coastal defense issue in his (still-confidential) testimony before the Lassiter Board.[7] It was this particular roles and missions issue that prompted Patrick to seek additional ships as targets for Air Service bombers. The successful sinking of more warships would credibly emphasize Patrick's Lassiter testimony about the offensive capabilities of the Air Service.

A major obstacle, though, was the U.S. Navy itself. Ever since the *Ostfriesland* sinking, and despite repeated urgings by Secretary Weeks, the Navy refused to provide any ships. Patrick pressed the issue in February 1923, and Weeks wrote to Navy Secretary Edwin Denby requesting the availability of "obsolete naval craft." Denby's answer of 20 February was a convoluted, slanted interpretation of the Washington Naval Treaty requirements for scrapping.[8] After several impatient proddings by Mitchell, Patrick pressed the issue again with Assistant Secretary Wainwright in early March 1923. This resulted in two 19 March meetings. Patrick had a morning session with Assistant Secretary of the Navy Theodore Roosevelt Jr., followed by an afternoon session with Adm. Samuel S. Robison, a member of the Navy Headquarters staff. Afterwards, Patrick retired to his office and wrote a lengthy, confidential memorandum summing up the discussions. While Roosevelt seemed accommodating, Admiral Robison was "not particularly inclined to aid in carrying out the bombing experiments."[9] Patrick commented, "After this inter-

Cadet Patrick in his first year at West Point Military Academy in 1882. Patrick, and classmate John J. Pershing, held the top two posts, respectively, as first and second captains of the Corps of Cadets for the class of 1886. (Courtesy United States Air Force)

Patrick as AEF Commander Gen. John J. Pershing's Chief of Air Service, AEF, in 1918. (Courtesy United States Air Force)

Above: Maj. Gen. Mason Patrick, Chief of Air Service, AEF, and Brig. Gen. Billy Mitchell, First Army Air Service Commander, participate in an awards ceremony in France, 1918. (Courtesy United States Air Force)

Left: Brig. Gen. Benny Foulois, who preceded Patrick as chief of the Air Service, AEF, with Gen. John J. Pershing, the AEF commander, in April 1918. Foulois and his large staff of 412 personnel descended en masse to augment the Air Service AEF staff assembled by Billy Mitchell. Problems with management led Pershing to later characterize the Air Service staff effort as "good men running around in circles," and to replace Foulois with Patrick in May 1918. Foulois would go on to head the Army Air Corps from 1931 to 1935. (Courtesy United States Air Force)

Above: Col. Henry H. "Hap" Arnold when he served as the War Department's chief, Signal Corps Aviation Division, in 1918. Handpicked by Patrick to be his Air Service Headquarters chief of information, Arnold was subsequently banished by Patrick to Fort Riley, Kansas, in 1925, for his role in attempting to lobby congressmen in violation of administrative orders. (Courtesy United States Air Force)

Below: Capt. Eddie Rickenbacker, famous AEF, Air Service ace, and friend of Billy Mitchell. Rickenbacker noted in 1921, after Chief of Staff General Pershing had once again asked his friend Mason Patrick to take over the Air Service, that "the appointment [of Patrick] is as sensible as making General Pershing admiral of the Swiss navy." (Courtesy United States Air Force)

Above: Patrick was instrumental in laying the groundwork for the strategic bombing mission, strongly supporting the implementation of a joint long-range bombing program with the Royal Air Force. Patrick appreciated the capabilities of long-range (strategic) bombing but was realistic about its limitations, given the current state of the art. (Courtesy United States Air Force)

Left: Patrick chose Col. Edgar Gorrell to oversee the writing of the AEF, Air Service history. Gorrell was a key contributor to the formulation of early Air Service long-range bombing doctrine. Although he returned to civilian life after the war, he and Patrick maintained their friendship and correspondence. (Courtesy United States Air Force)

Though Maj. Gen. Charles T. Menoher was a distinguished combat commander during World War I, as chief of the Army Air Service following the war he failed to keep Billy Mitchell in line. This subsequently led to his replacement by Mason Patrick as chief of the Army Air Service in October 1921. (Courtesy United States Air Force)

Having returned to the States in 1919 after serving in several roles at the Versailles Conference at Pershing's behest, Patrick reverted to his permanent rank of colonel. He was then brought back by Pershing in October 1921 to once again command the Air Service and, specifically, to control Billy Mitchell. (Courtesy United States Air Force)

Above: Patrick fully realized the importance of being a "flying chief" of the Air Service. Shown here with Billy Mitchell in 1924, Patrick was awarded his Junior Aviator wings in the summer of 1923 at age fifty-nine. Patrick still holds the record for being the oldest individual on active duty in the U.S. military to earn his wings. Patrick maintained his flight proficiency until he retired at age sixty-four in December 1927. (Courtesy United States Air Force)

Right: Maj. Bert Dargue, a member of Patrick's staff and at one time his pilot, taught the Air Service chief to fly. (Courtesy United States Air Force)

Capt. Ira C. Eaker, who served on Patrick's headquarters staff, went on to become Eighth Air Force commander and commander-in-chief of the Mediterranean Allied Air Forces during World War II. Patrick was directly responsible for retaining Eaker in the Air Service. Eaker was about to depart the Air Service in 1922 to attend law school when Patrick arranged for his transfer to Washington, D.C., to obtain his law degree while on active duty. Eaker went on to serve on Patrick's headquarters staff as assistant executive officer. (Courtesy United States Air Force)

Maj. Gen. M. M. Patrick, chief of the Air Corps, Lt. Lester J. Maitland, pilot of Fokker Trimotor on Hawaiian Flight, Brig. General J. E. Fechet, Asst. Chief of the Air Corps, and Lt. Albert F. Hegenberger, navigation officer on Hawaiian flight of June 1927. Patrick was a strong proponent of endurance flights that demonstrated the capabilities of Air Service aircraft and personel. (Courtesy United States Air Force)

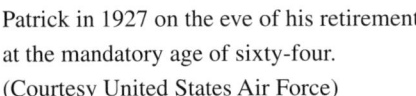

Patrick was a hands-on Air Service chief who studied all aspects of the aeronautical mission. Here, he poses with Lt. Comdr. Zachary Lansdowne (*left*) who, as commander, died in the ill-fated Navy dirigible *Shenandoah* disaster in September 1925. In part, it was the loss of the *Shenandoah* that prompted Billy Mitchell to call a press conference denouncing the administration's handling of the nation's defense program. (Courtesy United States Air Force)

Patrick in 1927 on the eve of his retirement at the mandatory age of sixty-four. (Courtesy United States Air Force)

view I am satisfied that the Navy are by no means enthusiastic over this work and that we may expect all sorts of obstacles placed in our way."[10] Patrick closed the memo by noting that the Air Service would press the Navy for the ships by 1 October, but "if there be any slip-up, it will still be practicable to do the work in the spring of 1924. This, however, I want kept to ourselves."[11]

Three months passed without a word from the Navy about ships. Mitchell's patience was long past exhausted; Patrick's patience was wearing thin.[12] Patrick finally sent an exasperated memorandum to Pershing at the end of June complaining about the Navy's "resistance."[13] Pershing broke the two ships loose from obstinate Navy moorings. Patrick did not get the *Dakota* and the *Delaware* that he had originally asked for; instead he ended up with the USS *New Jersey* and USS *Virginia,* and he immediately put Billy Mitchell in charge of this new round of bombing tests. Patrick gave Mitchell carte blanche as far as the tests were concerned. Unfortunately, to a degree, Congress and the War Department did not. Congress had specified no more than $50,000 for bombing tests against ships for fiscal year 1924. The other limitation, issued by the War Department, directed that the bombing be done from 10,000 feet.

While Pershing's assistance was a positive development, not so the unfortunate fact that President Calvin Coolidge had replaced President Warren G. Harding, who had died unexpectedly in the summer of 1923. Coolidge was no friend to military aviation. Just prior to the autumn 1923 bombing tests, in the midst of a discussion about greater economy in government, Coolidge startled some guests with the observation: "Now take those aviators, for instance. They just like to run around and burn a lot of gasoline. There's that Mitchell fellow. Why, he thinks nothing of flying in a government plane to Michigan to visit the girl he's engaged to marry."[14] Patrick was well aware of the reputation and rumors swirling about his second-in-command, and this was a critical time, not only because of Coolidge's temperment, but also because the Lassiter Board recommendations were in coordination between the Navy and War Departments.[15] Regardless, Patrick had faith in his second-in-command to accomplish the sinking of ships. Patrick stayed focused on the mission at hand and successfully deflected any criticism that the Air Service was "showboating" with the upcoming tests.

Concerns were aired that a new round of bomber versus ship trials would open old wounds between the services. The Air Service explained that "the proposed exercises are simply in the nature of training to increase the efficiency of the bombardment personnel of the Army Air Service."[16] That was true, decidedly so. This very public exercise would support Patrick's desire to showcase the offensive and independent capability of bombers. By extension, this would also complement Patrick's plea for five additional bombardment groups as requested before the Lassiter Board. In addition, Patrick saw this as an opportunity to get the coastal defense mission from the Navy.[17] Interestingly, some unique technical firsts went along with this bombing trial: the supercharged engine and a bombsight that was the antecedent of the Norden bombsight of World War II fame, both of which Patrick wholeheartedly endorsed.[18] With so much riding on the outcome of this bombing trial, both Patrick and Mitchell needed to win. Patrick, though, far outdistanced Mitchell in justifying and publicizing this event. In fact, Patrick was so successful at promoting the positive side of the Air Service in a nonconfrontational way that he made the cover of *Time* magazine in the summer of 1923.[19] During this period, Mitchell was shunted to the back pages.

The bombing of the *New Jersey* and *Virginia* took place on Wednesday, 5 September 1923, twenty miles off Cape Hatteras. As this was the first plane versus battleship trial since Mitchell's tour de force two years earlier, it was a pretty big show. Over three hundred assorted dignitaries watched the bombing effort from the deck of the Navy transport *St. Mihiel*. Acting Secretary of War Dwight Davis was there, accompanied by Pershing and an assortment of foreign military attachés, congressmen, and officers.[20]

What happened when Patrick attempted to give a briefing to the press is indicative of the fine line that not only he had to tread but that Pershing had to as well. Patrick had just begun his briefing when Pershing, unknown to Patrick, walked into the room. Patrick had only begun: "The differences between the Army and the Navy on bombs versus battleships . . .," when Pershing interrupted: "What General Patrick is telling you is that his Air Service is having some target practice." The briefing was abruptly over.[21] And by the end of the day, the bombing tests were over as well, and the two ships lay at the bottom.

Mitchell and Patrick had won. Pershing took off a great deal of the shine when he issued a statement on the results. It was drafted by the Air Service (and initialed by Patrick) and subsequently approved (eviscerated) by the Navy, making the point that nothing conclusive was proved by the sinking of obsolete warships. While Pershing put his weight behind obtaining the ships for the bombing trials, and did so successfully, it was also obvious that he did not wish to offend the Navy. Pershing went even further, banning any further official War Department publicity concerning the event.[22] Mitchell's report on the trials went into War Department confidential files and was not released, even by Mitchell. But Mitchell may have had more pressing things on his mind. He was to be remarried in October, followed by a nine-month combined honeymoon and inspection tour of the Pacific and Far East region arranged by Patrick.[23]

While Mitchell departed for the Far East and was concerned with things marital, Patrick got caught up in things martial as arguments between aviation enthusiasts and Navy protagonists took place in professional military journals.[24] The sharp argument that followed on the heels of the *Ostfriesland* bombing trial repeated itself with the latest round of plane against ship. This controversy played right into Patrick's hands. It provided a forum whereby he could almost seem to be a dispassionate commentator explaining both sides of this contentious issue. In a 6 October 1923 *Army and Navy Journal* article, Patrick provided what was calming heavy oil on troubled waters. Taking his lead from Pershing, Patrick wrote, "The Air Service does not for a moment assume to say that the battleship or any other component of the naval establishment is obsolete."[25] Mitchell, on the other hand, believed they were.[26]

Patrick was a firm supporter of not only combined arms, but of joint warfare as well. The speeches and lectures that Patrick presented from November 1923 to early March 1924 were prime examples of his Air Service philosophy and his national defense philosophy, which he did not mind sharing. Patrick was a good impromptu speaker who spoke with few notes. His technique consisted of legal-sized sheets of paper upon which were listed his main talking points, sometimes with corresponding "lantern slides," or "plates."[27] Regardless whether the topic was technical or more prosaic, Patrick handled it well.

As a professionally schooled engineer, Patrick was attracted to the technical side of aviation, and he was at ease in the company of those versed in the technical side of the trade. He was so at ease, in fact, that he was invited by the dean of the Engineering Department at Harvard University to deliver a detailed lecture to the Harvard Engineering Society (which in hard copy ran for twenty-five single-spaced pages) with just four pages of abbreviated notes, much of which was engineering background data.[28] Patrick's public presentations were not limited strictly to things aeronautical. Having made many acquaintances within the engineering community during his distinguished career, Patrick received innumerable invitations to speak before a variety of professional engineering groups and managed to fit many in.[29] But more important to him than the professional civilian associations were the Professional Military Education (PME) schools.

Patrick had a full itinerary to speak before the members of the Army establishment, and they were anxious to hear him out. He spoke once or twice a year to the Army War College classes in Washington, D.C. His first speech, in March 1922, was a bland description of the "design, procurement, production, and maintenance" problems of the Air Service, a topic requested by the Army War College commandant.[30] Patrick's second 1922 speech before the Army War College was much different, keyed to the issue of air supremacy.[31]

Patrick's 1923 invitation to speak at the Army War College came directly from the commandant, Maj. Gen. Hanson E. Ely: "In view of the increasing importance of aeronautics in land and sea operations, it is desired to keep the War College faculty and students informed of developments in theory and practice."[32] After noting that each faculty division expressed a keen desire for an expanded segment on the Air Service, Ely stated, "The purposes of the college would best be served, it is believed, by setting aside a morning to be devoted entirely to an orientation in the status of our Air Service."[33] Ultimately, both Patrick and Mitchell presented lectures. Patrick's topic covered the development of "military aeronautics" among the major powers, while Mitchell discussed "Tactics of the Air Forces in War." These presentations were followed by a "general conference" to include faculty and students.[34] It is noteworthy that a large portion of Patrick's lecture was devoted to the importance of the commercial manufacturing base to the military air effort. Mitchell did not pontificate on the need for Air Service independence. His script did not deviate from the topic at hand.

Patrick had been fine-tuning Air Service doctrine since his first day on the job when he spoke with his budget officer, Capt. Roger Volandt. The direct relationship and correlation between the roles and missions of a service arm and its budget was something that Patrick was no stranger to. The economy environment within which Patrick struggled to maintain his service forced him to become a strong advocate for doctrinal change, just like Billy Mitchell, but not in the same mold.[35]

For the remainder of 1923 Patrick continued to spread the Air Service word in person and the written word. Just as Mitchell used the press and professional journals to advance aviation interests, so did Patrick. Patrick's emphasis, though, was primarily on "in-house" publications: *U.S. Air Service, Army and Navy Journal,* and *Military Engineer.* Fresh from his stellar performance before the Lassiter Board, Patrick used these publications to "reach large sections of the Army population or . . . the men who would be making Army doctrine in the future."[36] The presentations before the War College in November 1923 were significant not so much for what was said but because of the expanded interest shown by that institution compared to the previous year's presentation. By contrast, a March 1924 lecture at the General Service Schools at Fort Leavenworth was notable because of its topic: aerospace doctrine. This was a seminal speech. The General Service Schools trained officers destined for the General Staff, who, in turn, would become the future leaders of the Army. For many, their job would be the articulation and implementation of Army doctrine. On 27 March 1924 Patrick presented the lecture to the assembled General Staff officers in training. At this time the draft of "Fundamental Conceptions of the Air Service" had not as yet been accepted by the General Staff as a Training Regulation. It was still in coordination. But one institution that did review it was the General Service Schools. When Patrick began his speech, he made light of this fact, noting his unfortunate choice of words in one part of the document and, at the same time, putting everyone at ease:

> To say that [the Training Regulation] did not meet with unqualified approval is putting the case rather mildly. A study of the comments offered indicates that the text conveyed to the reviewers the erroneous idea that the Air Service intended in the future to fight all wars by itself and that in its opinion the remaining branches of the service could safely stay at home.... I can say most decidedly that it was not the idea which it was intended to convey.[37]

Patrick assured the audience, in no uncertain terms, that "the mission of the Air Service is to assist the ground forces to gain strategical and tactical successes."[38] He went to great pains to alleviate fears over the issue of the "Air Force" operating independently of the ground forces. He admitted that the Air Force would indeed operate independently, and do so sometimes far afield of the current theater of operations, but it would be done "in carrying out the common mission." All activities, both Air Service and Air Force, would be "under the immediate control of the Army commander." Patrick was sharply critical of those who would narrowly define the Air Service's auxiliary status to being tied directly to the immediate theater of operations and ground support. His rejoinder to that point of view was blunt: "we believe that this is a misconception of the possibilities of the Air Service and we cannot be tied down by any such doctrine."[39] Further on, Patrick described several fundamental items of military air doctrine: "that there is no adequate defense against an air attack except an Air Force of our own," and "that the command of the air, air supremacy, means command of the surface, whether it be on land or sea."[40] Patrick was indeed visionary when he stated,

> The mobility of an Air Force is one of its most important characteristics, but to be effective it must be possible to send with it such supplies, spare parts, ammunition and the like that will be needed in active operations. The day will no doubt come when such a force will be but little dependent upon ground transportation.[41]

A Shot Across the Navy's Bow

Doctrinally, Patrick had come a long way. He was not only an advocate for an independent Air Service mission outside of the immediate theater of operations, but he also foresaw the future expeditionary promise of an air force. But in the latter half of 1923 Patrick looked closer to home to expand the mission of the Air Service. By this time he had already succeeded in developing a formal operational doctrine that would ultimately be codified as Training Regulation 440-15. What he planned to do next was expand the Air Service mission to encompass coastal defense, not just only the defense of the immediate coast, but out to a distance of 200 miles.

There was a very practical reason for this initiative. The defense of the continental United States would always have priority, and Patrick believed the Air Service deserved a role in that defense. The historical division of labor for this mission was simple: the Navy would fight on the high seas to turn back a foreign invader, and the Army would fight on land if the Navy failed and invasion followed. The advent of the airplane, and specifically the bombardment aircraft, changed that simplistic arrangement. Bombardment aircraft, whose effective range was increasing almost yearly, offered the Army, in the form of the Air Service, the opportunity to literally extend its role 200 miles out to sea or more. Mitchell, when he was pushing for this role in 1919, did not even go so far as to specify a distance. Patrick not only supported this Mitchell initiative, but he also had faith that the bombers could succeed in doing this historically U.S. Navy mission encompassing an appreciable area.

The airplane's part in this particular aspect of national defense was the topic of several boards of inquiry in 1917, 1919, and 1920. All of them failed to institute a concrete division of labor, and the Navy was loath to give up the hundreds of land-based scout aircraft dedicated to the ocean reconnaissance mission.

The crux of the problem was War Department General Order No. 4, issued in 1920, covering the functions of Army, Navy, and Marine aircraft. This order gave the Navy free reign over the coastal area, both land and sea, both from "mobile floating bases" (aircraft carriers) and naval air stations on shore. In fact, Army aircraft were much more limited in their mission by the General Order. Army aircraft functions were limited to acting against enemy aircraft in defense of shore establishments and against enemy vessels engaged in attacks on the coast.[42]

The Navy had this mission courtesy of General Order No. 4, and because the General Staff was satisfied with the arrangement and ignored Patrick's protests on the issue when he assumed command. But Patrick had other reasons to be sensitive to the Navy's role for land-based aircraft; they were encroaching on his territory. The Navy had been attempting to supplant the Air Service at Rockwell Field near San Diego ever since Patrick took over in 1921. Patrick was kept fully informed of the situation by several individuals, and one in particular, Hap Arnold. As the Rockwell Field commander between 1922 and 1924, Major Arnold wrote to Patrick complaining about and documenting the Navy's expansion and encroachment.[43]

By the end of 1924 absolutely no progress had been made on the coastal defense issue. The War Department failed to respond to Patrick's protests for two reasons. The first issue was Pershing's sensitivity to maintaining good relations with the Navy. The second issue was the fact that few individuals in the War Department, not to mention the public at large, believed at this time there was any credible threat of air attack from across the seas.[44] Patrick left this issue unresolved, but not forgotten. Patrick himself had a reason for not pushing too hard for this particular issue, because another issue dear to his heart was coming to a head. Patrick was leading a personal crusade against one of his own divisions, McCook Field, as well as crusades in support of the aviation manufacturing industry and civil aviation development.

9. Patrick's Search for Economy and Efficiency

Elimination of the McCook Field Engineering Division

Shortly after Patrick took command, he moved to consolidate the Air Service storage depots into five regional centers. Next, all pilot training was centralized at San Antonio, Texas. Technical training was then consolidated. Excess Air Service properties were sold, and a handful were deactivated.[1] Patrick was inspecting, evaluating, and redistributing the assets of the entire Air Service.

Patrick also evaluated, as he termed it, "the so-called Engineering Division of the Air Service." He found that the "work was being carried out almost independently and without any considerable measure by the office of the Chief of the Air Service."[2] This particular issue, the status of the McCook Field Engineering Division, highlights a distinct difference between the way Patrick and Mitchell approached their jobs. Whereas Mitchell was concerned with the technical aspect of flight from a viewpoint of aeronautical improvements, Patrick was more concerned with the "value received" aspect of the Air Service's research and development program.[3] Mitchell maintained a very close relationship with the Engineering Division, almost running it as a private fiefdom.[4] He was aided in this endeavor by his close friendship with the Dayton commander, Col. Thurman H. Bane. Obviously, this Mitchell/Dayton connection did not go unnoticed, or unappreciated regarding its control aspects, by the chief of the Air Service. But Patrick's problem, in particular, centered not so much on one of Mitchell's principalities, but instead on the actions of the Engineering Division's Aircraft Design and Production Office. Patrick was concerned that funds the Air Service spent on the development of aircraft at McCook Field could be better utilized procuring aircraft that were designed, fabricated, and tested by commercial aircraft manufacturers. Patrick voiced this theme as early as 1921 but repeatedly touched on it beginning in 1923.

Patrick had another concern that was intimately related to what the Engineering Division was doing. Surplus aircraft were a long-standing problem; the Air Service was directed by Congress to use war surplus aircraft until they became unservicable. This was not a small concern that would go away in a few years. Patrick had already been called on the congressional carpet by the Frear Committee at the close of the war for destroying what many erroneously thought were perfectly good aircraft.[5] When Patrick took over the Air Service again, he was faced with a glut of 220 different types of obsolete and obsolescent aircraft: a total of 3,100 aircraft.[6] While flight safety was a major concern due to the condition of these rapidly aging aircraft, Patrick could not by law arbitrarily get rid of the aircraft. But he did. Between October 1921 and June 1922 he scrapped 920 aircraft, using the same cannibalization and auction process he had used in Europe.[7] In the fiscal year ending June 1923, only 746 war surplus aircraft were on the books. When asked why he did not sell the planes outright, he replied that the government should have no part of providing the public with anything of an unserviceable nature for monetary gain, knowing that it would probably result in loss of life and property.[8] And if anyone questioned Patrick about the serviceabilty of the scrapped aircraft, Patrick had only to point out that the planes his pilots were flying were not much better. During 1921 330 crashes had taken the lives of sixty-nine Air Service officers.[9] The odds were one in ten that a pilot or observer would die if things did not change for the better. By the summer of 1922 the Air Service classified only 910 aircraft as airworthy. Despite Patrick's efforts, things were getting worse.[10] Using the 1921 calendar year statistics as a starting point, if one compares the fatalities for the next two years, fatal crashes due to engine trouble increased by a factor of two while fatalities due to structural problems increased by a factor of ten.[11]

In Patrick's mind, the U.S. government was responsible because they did not give the aircraft manufacturers a fair deal when it came to competing for contracts or protecting their research and development risk. The government was ignoring the health of the aviation industry at its own peril. Patrick was a firm believer in the vitality and financial health of a commercial and civil aviation infrastructure and manufacturing base. He set out to make that a reality. Given his belief in an offensive role for the Air Service on D-day, it was obvious to him that an aviation industry had to be "in being" prior to a conflict. To play catch-up after the start of a war would almost guarantee failure. This was based on Patrick's experience during World War I, when the average service life of a single-seat fighter was six weeks.[12]

Not that the government was the only guilty party. The aviation industry did not cooperate either, resisting industry standardization and covetous of their few contracts.[13] With no government subsidy in any form, the aircraft industry was left to its own fate. Patrick was determined to assist the manufacturers and, in doing so, assist the Air Service as well.

Patrick began by addressing the part of the problem he could deal with directly, the competition given the aircraft industry by his own Engineering Division. Granted, there were aspects of the Engineering Division's work that Patrick termed "of extreme importance and being well done." It was the office of airplane design, in particular, that Patrick was philosophically and financially opposed to. When Patrick took over as chief, he inherited an aircraft procurement system based on long-standing regulations that gave the Engineering Division absolute control over which aircraft designs were sent forward to Air Service headquarters with a recommendation for production. After the war there was a further tightening of control due to a congressional and Air Service backlash from the contract abuses of World War I.[14]

Patrick did not have a problem with the Engineering Division evaluating the designs of commercial manufacturers. In time he even strengthened this aspect of its charter. What he did have a problem with was that the in-house Aircraft Design Office was in direct competition with commercial manufacturers and was thus stifling competition. Patrick did not come to this conclusion independently, and he forthrightly admitted it. He was lobbied hard by the aircraft manufacturers to change the rules. As Patrick described the situation:

There was the claim on the part of the manufacturers that the engineers at the Division always preferred, and gave preference to, designs which had originated with them. In other words, outside agencies were thus brought directly into competition with Government employees doing designing. It was but human nature for these government employees to give the product of their own skill the preference.

... At this same Engineering Division aircraft were being built . . . and again the manufacturers complained that this was undue interference with their enterprises.

... I became convinced that the contentions of the manufacturers were valid and that it was essential to change radically the Engineering Divisions practice.[15]

Patrick heard and read a great deal about this problem from the manufacturers' perspective, but he did not immediately do anything of substance to correct the situation.[16]

If anything, Patrick was thorough.[17] He analyzed a problem based not on the strength of the advocate's argument, but on the strength of the entire picture as it affected the Air Service, which sometimes took an extended period. As he himself noted, "It took time to study these matters, to decide just what was proper, the best course to follow."[18] So it was with the Engineering Division issue. To prove just how much time Patrick devoted to this issue, one need only look at his office diary. The first time that Patrick discussed this issue was on 7 October 1921, after only two days on the job, when he spoke with Captain Volandt, his budget officer. Part of the discussion concerned "finances transferred to the field."[19] When Patrick saw the budget numbers, he was doubtless impressed with how much was allotted to two particular budget items for the Engineering Division at McCook Field. "Experimental research, plus engineering and development" received over $2 million; "Production of new aircraft, engines, and accessories" received over $10 million.[20] The "production of new aircraft, engines, and accessories" at McCook Field accounted for a third of the yearly Air Service budget.

Patrick bided his time on this issue for a variety of reasons, most of which were political. It was not until Secretary Weeks approved the Lassiter Board report, which contained a proposal for significant aircraft augmentation, that Patrick actually issued orders to reorganize the McCook engineering organization. But by then he had all his facts down and knew precisely where he wanted to go.

In early February 1923 Patrick was busy preparing for his presentation before the Lassiter Board in March. As part of that preparation, he sent a memo to his staff: "In considering industrial preparedness for war, a study must be made of the aeronautical industry."[21] Sometime after 20 February Patrick was given a two-page memorandum in response to his tasking. He used this information as the basis for his testimony before the Lassiter Committee to argue the point that the aircraft industry could never support wartime requirements.[22]

Due to Patrick's foresight and his staff's research, he was able to use this information to great effect. He was able to provide to the Lassiter Board, and subsequently to many audiences, the absolute inability of the United States' aeronautical industry to support current war plans. Excluding the year 1919 because of skewed figures due to war contracts, Patrick showed that for the years 1920 through 1922 the aviation industry produced less than 2 percent of the required wartime production rate.[23] And with each passing fiscal quarter, more aviation manufacturers were going out of business. Secretary of War Weeks agreed with Patrick on this vital issue: "The aircraft industry in the United States is entirely inadequate to meet peace and war time requirements. It is rapidly diminishing under present conditions and will soon practically disappear. It depends for its existence almost wholly upon orders placed by governmental services."[24] Patrick also made his point about the lack of peacetime support with the statistics on Air Service fatalities due to rapidly aging aircraft.

These points were made part of speeches that Patrick's office drafted for Dwight Davis, the assistant secretary of war, speeches that set forth official War Department policy regarding the relationship between the commercial aviation manufacturing base and military aviation preparedness. Typical of these presentations is one Davis gave before the St. Louis Aeronautical Corporation in October 1923, in which Patrick and his staff incorporated the major points of the approved Lassiter Report:

The threatened extinction of the aeroplane industry is a national peril and must be averted. It is our national policy to use existing commercial facilities to the greatest possible extent, rather than to maintain large military establishments in time of peace. We should adopt a ten year programme of building planes, not as an aggressive measure or in competition with other nations, but merely to meet the deficit in planes needed for purely defensive and training purposes.[25]

Patrick hammered on this theme as well when he spoke and wrote for domestic and foreign military and civilian audiences.[26]

With Secretary Weeks, Assistant Secretary Davis, the General Staff, and supporting statistics behind him, Patrick made his move to eliminate the aviation design and construction responsibilities of the Engineering Division at McCook Field. In late November 1923 Patrick directed that Maj. L. W. McIntosh, the chief of the Engineering Division since July 1922, be prepared "to defend your position in reference to the amount of Air Service funds allocated annually to the Engineering Division."[27] Major McIntosh finally responded on 27 May 1924.

Some time ago you intimated that there was in your mind some uncertainty as to whether or not the Air Service was getting "value received" for the [Engineering Division] money thus expended.

. . . You directed me to consider the matter and this memorandum embodies as complete a statement as can be made from the Engineering Division. I am calling your attention to the work of the Division in detail, and the necessity for that work, in order that you may be properly informed of the consequences which would befall the proper equipping of the Air Service, should you restrict the activities of the Division.[28]

The response from McIntosh did not dissuade Patrick from his task at hand. In fact, the dire "consequences" in the condescending response from Major McIntosh doubtless made Patrick move even faster on this issue. McIntosh was relieved of his position the following month and replaced by Maj. John F. Curry.[29] Patrick directed that henceforth the Engineering Division would not design or construct aircraft. It was directed to be strictly responsible for the test and evaluation of aircraft prototypes forwarded by commercial manufacturers and the development and testing of military aviation equipment.

Patrick's direction to the Engineering Division to emphasize the "development and testing of military aviation equipment" may have been due in part to two very personal incidents that persuaded Patrick of the need for additional work in this area. The Engineering Division had developed a urinary relief tube assembly with a tiny pump system that on one occasion, while in flight, reversed itself and in the process drenched Patrick. In another incident, on a cross-country flight, Patrick remained "painfully attached" to the relief tube assembly until rescued by ground crewmen.[30] In retrospect, Major McIntosh may have gotten his revenge on the man who fired him.

Soon, the flurry of Patrick-initiated activity within the Engineering Division drew the attention of Congress. When Rep. R. W. Ireland of the House Appropriations Committee got wind of what Patrick was doing, he called and asked for an explanation. Patrick replied:

As Chief of Air Service it has been my policy not to permit agencies under my control to compete with the airplane manufacturing industry. It is my earnest desire to foster and keep in efficient operation the airplane industry of this country. . . . McCook Field is simply a testing and developing plant for aeronautical equipment.[31]

When called for, prototypes would be built by manufacturers to the mission specifications provided by the Air Service, through the Engineering Division. But the aviation companies also had full freedom to submit any prototype for consideration based on their indigenous research, and Patrick had full faith that outstanding designs would be submitted.[32] What made Patrick so sure that the aircraft industry could supply sophisticated aircraft designs? He and much of the American public had seen the results in several highly publicized air races every year since the war ended. The latest designs were pitted against each other at these competitions, the planes many times flown by Army and Navy pilots. Patrick also knew many of the individuals involved with the aviation companies. Others he knew through professional associations. Patrick may have known the quality of the planes, and the quality of the men he was dealing with, but in the end it really came down to money.

Patrick knew that further budget cuts were on the horizon from economy-minded Congresses, not to mention the bottom-line business attitude of President Coolidge. In the 1924 presidential campaign Coolidge and the Republicans accused the previous Democratic administration of "muddling inefficiency" in running the War Department and wanted to relieve the public of "the burden of maintaining a large standing army."[33]

Indeed, the budget was reduced. But the budget reduction was almost entirely offset by Patrick's elimination of the Engineering Production Office. Although that portion of the budget was cut by $3 million in 1924, the remaining $7 million was diverted to the research, development, and production of all aviation items unique to military requirements such as armaments, navigation aids, and bombsights.[34] Patrick knew that his action was the right one. His study of the problem revealed that between April 1917 and November 1922 McCook Field actually produced only twenty-seven airplanes.[35] To Patrick's mind, the return was not worth the investment. Following the McCook Field Engineering Division reorganization, Patrick admitted that there were bitter feelings "but ultimately the conditions were radically changed; the feeling [of the manufacturers] toward the Engineering Division grew better and better."[36]

Patrick's reorganization encouraged a number of new aircraft designs to pour into McCook Field. Patrick timed this to coincide perfectly with the return in late 1923 and early 1924 of the almost $1.5 million in procurement funds that he voluntarily returned to Congress in 1922. Patrick had calculated that this sum would bridge the gap until formal action was taken on the Lassiter program recommendations. But Patrick took no chances on a quick approval of the Lassiter results. He submitted a request of $25 million for new aircraft procurement for the 1925 fiscal year.[37] Ultimately this was cut in half, but his long-term planning paid off.[38] Patrick reversed Air Service funding that had been going downhill since 1921. His initiatives during the 1922–24 period resulted in an annual increase in funding in fiscal year 1925, which continued up through 1931.[39]

Without government competition, the best new aircraft designs with the latest technology were submitted by assorted manufacturers in the latter half of 1924 and early 1925. These designs, in turn, were evaluated by McCook Field. When the choices were made, Patrick had the money to write contracts for the best aircraft (200 in all) then available to accomplish the new missions of the "Air Service" and the "Air Force."[40]

A good example of the benefits of this new procedure is evident in the story of the GA-1 ground-attack plane. In 1920 the McCook Engineering Division undertook development of the GA-1, an armored triplane using two Liberty engines, developed specifically for low-level bombing and strafing. Patrick approved of the GA-1 project. He was strongly in favor of the ground attack or "ground strafing" mission as he made clear in his *Final Report* to Pershing following the war.[41] When the new GA-1 planes arrived for the 3rd Attack Group at Kelly Field, Texas, their performance was so poor that the unit resorted to using modified war-vintage DH-4Bs as attack planes instead.[42] This performance reflected badly on the Engineering Division. The Air Service ultimately acquired an acceptable attack aircraft in 1927 when it purchased the Curtiss A-3. Patrick's elimination of the McCook Engineering Production office resulted in many more superior aircraft designs being procured just as the 1926 Air Corps Act and its associated five-year acquisition program took effect.

Patrick's initiative was successful in three other ways. He reorganized the Air Service for economical streamlined operation. This initiative resulted in more money for aircraft acquisition. Finally, Patrick eliminated direct government competition with aircraft manufacturers. He had inherited an archaic and highly proscribed contracting process. Fixing it was his next challenge.

To the Lowest Bidder Go the Spoils

Patrick was faced with a contracting system that allowed the Air Service little leeway in the awarding of contracts.[43] Government contracting was again based on peacetime competitive bidding; the lowest bid almost invariably won out, with little thought given to the capability of the winning bidder to complete the contract successfully. The only loopholes available to the Air Service were if the contractor was the sole source of the item or if the item was patented. But here again, Patrick was stymied. The few aircraft manufacturers who were still in business jumped at every contract bid, and, according to law, an airplane could not be patented simply because a new aircraft was not considered new technology worthy of a patent.

Patrick wanted to establish a better contracting system with two initiatives: (1) he wanted the best qualified companies to receive government contracts to ensure their continued existence; and (2) he strove to eliminate the requirement for the aircraft designer (the company) to sell the design rights of its aircraft to the government. This latter requirement, plus the use of competitive bidding, ensured inferior aircraft if the contract was completed at all.

The aircraft contracting (acquisition) process of the early 1920s guaranteed poor results. An aviation manufacturing company, with no government subsidy, created a new aircraft design and constructed a prototype. Perhaps the design/prototype was created to meet a specific Air Service requirement. The plans and prototype were forwarded to the Engineering Division at McCook Field for testing and evaluation (T&E). Once the prototype was approved by McCook and Air Service headquarters, the company who produced the design and the prototype was required to sell the design, prototype, and rights thereto to the Air Service. At that point the Air Service requested bids. The Air Service, by law, then accepted the lowest bid. On the surface, the competition process worked. Or did it? At this point, the original designing company, knowing the associated construction problems, and with design costs to recoup, would almost invariably submit a higher bid than the competing companies that had no previous sunken costs to recover. But if the original design company chose to make a financially suicidal low bid to ensure being awarded the contract, it risked bankruptcy, which often happened. On the other hand, if another company got the contract, problems usually arose because this company had no experience with the design and no appreciation of its manufacturing intricacies. Costs increased, companies went bankrupt, the Air

Service got its planes late or not at all, and the original design/manufacturing aviation firm gained nothing.

Patrick attacked this issue by successfully lobbying Assistant Secretary of War Dwight Davis, who supervised all War Department procurement, to change the rule concerning proprietary design rights. Davis eventually ruled that the government would "recognize . . . the principle of proprietary design rights." Thus Patrick could invoke a sole-source requirement, due to the patent on the aircraft design, and be assured that the originating company would be well positioned to provide a good product.

With the Patrick-directed closure of the McCook Engineering Division Production Office, plus the Patrick-inspired proprietary design rights initiative, one could accuse him of being "in bed" with the aviation manufacturing companies. The Air Service was one of the most investigated federal agencies of the 1920s, and Patrick was well aware of it. Five days after Patrick reported for duty as Air Service chief, he was visited by a Mr. Ronald Scaife from the Department of Justice. Scaife called Patrick's "attention to the fact that there had been a number of investigations in the Air Service—the Hughes Investigation, Senate and House [investigations], and numerous [issues of] improper conduct, that is, affairs of the Air Service conducted improperly—waste, inefficiency, and graft."[44] Scaife went on to say he had been directed by the Justice Department to investigate any and all things pertaining to any irregularities and asked for Patrick's assistance. Patrick directed his staff to give Scaife their full cooperation. Patrick himself noted that "[t]he Air Service . . . since we entered the World War has probably been the most investigated activity ever carried on by the United States."[45] There was never any evidence, or accusations, or even hint that Patrick's actions and motives were anything but legal and honorable.

Patrick's Support for Civil Aviation Development

Patrick's third initiative during this period with regard to economy and efficiency was an uncompromising support for civil aviation. Patrick was not a recent convert to the importance of civil aviation. As early as 1919, in testimony before Congress, he explicitly recommended federal control of aviation. He also included this suggestion in his lengthy September 1919 memo to General Pershing, when the AEF commander was asked by Congress for his views on postwar aviation.

Patrick's August 1919 testimony before various congressional committees was not the result of personal whim, but was based on a thorough grounding in commercial aviation issues with which he had dealt in his postwar role. As a member of a Versailles Peace Conference subcommittee on postwar aviation matters, Patrick was well aware of the civil aviation situation in Europe. While serving on the treaty subcommittee, Patrick made a diary notation concerning the great promise that commercial air transportation holds for the world.[46] He was not thinking just of the United States, he was thinking of the future possibilities of international flight.[47]

Patrick preached three basic tenets concerning the importance of a nation's air power: a strong domestic aviation production industry; a well-developed commercial aviation transportation system; and a capable military air arm in the event of war, especially one part, the "air force," which would be capable of immediate response. Patrick presented these notions in one of his earliest speeches after becoming Air Service chief. In "The Relation of Commercial Aeronautics to National Defense," Patrick began his speech by stating, "The success of Commercial Aeronautics is a military necessity."[48] He went on to emphasize the requirements for a strong aviation manufacturing base and a military air arm ready to respond immediately to aggression. But the overall theme of this speech, plus dozens more in the years to come, was Patrick's conviction that without both a strong aviation manufacturing industry and a civil aviation transportation system in place, the United States would be courting disaster if war should come again.

Surprisingly, Patrick had a powerful ally in this quest. President Harding's 12 April 1921 annual "President's Special Message to Congress" directed the Army Air Service "in cooperation with other agencies of the Government in the establishment of national transcontinental airways, and, in cooperation with the states, in the establishment of local airdromes and landing fields."[49]

Prior to this presidential directive, the Air Service itself had been engaged in the establishment of a modest scheduled airway system between several of its airfields and encouraged municipalities across the country to construct airfields for civil use. Of course, these would also be very advantageous for military purposes. In addition, a model airway program was inaugurated at Bolling Field on 12 February 1921, connecting Washington, D.C., with Dayton, Ohio. Due to lack of funding and support personnel, both programs were soon discontinued by General Menoher, and the Airways Section, which was originally established in September 1920, was disbanded.[50]

Although President Harding in his April 1921 message emphasized the importance of this issue, it did not translate into financial support. Between the time of Harding's message to Congress in April and Menoher's departure from the Air Service in October, absolutely nothing more was accomplished in the Air Service with regard to this mission until Patrick took over. Only then, with Patrick's 1 December 1921 reorganization of the Air Service headquarters, was the Airways Section reestablished.[51] It was obvious that the Airways Section under Menoher was given neither the support it needed nor the publicity. Shortly after Patrick formally reestablished the section in December, the Airways Section received several inquiries from commercial aviation concerns requesting more information about its responsibilities and course of action.[52] These aviation concerns had to wait a few months for a definitive answer because even though Patrick had directed Lt. Col. James Fechet, the chief of the Training and War Plans Division, to reestablish the Airways program, much needed to be done.

Fechet explained in detail in an April 1922 memorandum to Patrick just how formidable the task would be. In short, the Air Service was tasked by the War Department with all aspects of commercial aviation development, to include recommendations on legislation, the development and operation of airways, associated airdromes, and aerial mapping.[53] Fechet, not one to mince words, got to the heart of the problem: "In considering the duties assigned to this section . . . it should be considered as a small nucleus of what will later develop into an agency of the government . . . for control of civil aviation."[54] Fechet went on to ask for "an enlargement" of the Airways Section and support by the War Department to obtain the much-needed increases in personnel.

Patrick knew the importance of establishing a strong commercial aviation base. Menoher, by contrast, was fiscally shortsighted. Patrick was quite practical and farsighted. Patrick, not getting any additional support from the War Department, simply took money from other activities to sustain and broaden the work of the Airways Section, which he considered of paramount importance in his grand scheme of commercial aviation support.[55]

Patrick had a presidential directive to do as he pleased regarding the development of commercial aviation in the nation. Unlike Menoher, Patrick took a personal interest in the Airways Section and ordered an expansion of the Model Airway system. This organizational feat was the basis for the U.S. Air Mail System, which by 1924 could boast of one-day service between the coasts and an enviable safety record.[56] Patrick was innovative. With the continued proven success of the U.S. Air Mail System, he proposed a better way of doing business. Patrick's innovation was to utilize the same air-mail route system for air-freight purposes. By publishing figures on the delivery times of certain items of commerce, and especially interest-bearing financial instruments, Patrick held out the possibility that businessmen and financial institutions could save a significant sum by utilizing air transport.[57] To prove to the public and the business community just how reliable air transport could be, Patrick pointed to the safety record of not only the U.S. Air Mail System but also his Model Airway system. By 1923 the Model Airway had operated continuously without accident since 12 February 1921 and was averaging 7,000 miles per week.[58]

Utilizing the Air Service Airways and Information Sections, Patrick pumped out press releases describing the progress being made in commercial aviation. These press releases regularly presented the public with statistics on miles flown, numbers of air passengers, pounds of air mail delivered, and volume of air goods traffic.⁵⁹ Within these press releases, Patrick also included comparative statistics of European and American commercial air progress. The dismal figures for the American effort probably spoke volumes without any further commentary.

To further assist the smooth operation of the Air Mail System, the Model Airway, and, by extension, civil aviation, in March 1923 the Air Service Information Section began publishing a quarterly pamphlet, "Airways and Landing Facilities," which listed all surveyed landing fields.⁶⁰ Patrick took this one step further. Using his great influence with his original service arm, Patrick had the Corps of Engineers prepare a map of the United States detailing the location of those fields and the proposed airways between them.⁶¹

Although these initiatives were incremental steps in the right direction, by early 1924 Patrick was frustrated with the overall poor status of America's efforts in the civil aviation industry. The field was in desperate need of federal regulation. He summed up his frustration when he said, "It is manifestly impossible for the War or Navy Departments or a Separate Air Service to cope with such a situation as a side line to National Defense."⁶²

To Patrick's mind, the three legs of the aviation triangle—military air, commercial air, and the aviation manufacturing base—were completely unbalanced. He was attempting to make them mutually supporting. In the early 1920s the military aviation leg was the strongest of the three, and that was not saying much. There were assorted reasons for this: some practical, some economic, and others based in hidebound tradition. On the practical side, both the public and the business community saw little value in an unproven transportation system of seemingly limited utility. This was particularly so in the United States. The Europeans, on the other hand, had developed a lead in both military and civil aviation prior to the war and still maintained it. Patrick witnessed that advantage first-hand after the war, and with that first-hand knowledge he also knew why the Europeans were ahead of America in this endeavor. To a great extent, it came down to government subsidies. The British, French, and Italians all gave generous subsidies to their aviation industries. The Harding and Coolidge administrations would have none of that. Patrick presented a levelheaded rebuttal: "There is not a single method of transportation that this country knows now that has not been subsidized, either directly or indirectly. It has been so with railroad land grants. The government dredges rivers and harbors and roads are constructed to subsidize automobile traffic."⁶³ Patrick took his message on the road, and his audience was wide-ranging: executive clubs, engineers, civic groups, business councils, and varied professional organizations.

Patrick gained some support from another high-ranking representative of the War Department. Several speeches along these very same lines were drafted by Patrick's office (usually by Lt. James E. Van Zandt) and given to Secretary of

War Dwight Davis at his request.[64] When requesting the preparation of an article, though, Davis made plain his approach to the topic when he asked for, "a moderate article on Commercial Aviation and the National Defense."[65] Secretary Davis eventually took Patrick's commercial aviation theme to heart, albeit with moderation. Davis supported Patrick's elimination of the Production Office at the Engineering Division and also called for subsidizing the aircraft industry and civil aviation.[66] Davis laid the blame at the feet of the government, but gingerly:

With the best of intentions . . . the government is severely handicapped in its efforts to create and sustain an aeronautical industry. . . . If the United States, in the days before the advent of commercial aviation, is to possess the basic aircraft industry that its national security requires, some solution of the present difficulty must soon be discovered.[67]

In 1924, at Patrick's urging, Secretary of War Davis officially recognized the critical fact that civil aviation was nonexistent. Patrick knew its critical importance in 1919.[68]

Patrick was historically attuned to the mood of the nation. He knew that there would never be support in the overall population or in Congress for an air force large enough to overwhelm or blunt an enemy in the opening days of a war. "We shall never maintain a large Army nor will it be possible to have an Air Force of sufficient size to protect us in case we are unfortunately engaged in a war against a formidable enemy."[69] The only hope to even approach that capability rested with demand for aircraft from a burgeoning domestic civil aeronautics industry. According to Patrick, this achieved two ends. It would vastly stimulate the aircraft production base and, second, it would create a large, ready-made pool of trained reserve pilots in case of war, individuals whose training would be much enhanced by their previous occupation. This would both save money and leave the country prepared in case of war.

Patrick's views on the subject were made clear to the father of one of his pilots who had just graduated with honors from the flying course at Kelly Field. The father had written to Patrick asking for advice regarding his son's future: should his son make the Air Service a career or should he obtain a position with the U.S. Post Office Air Mail System. Patrick replied: "I am thoroughly satisfied that air transportation is coming into being and this more rapidly than we now think probable. . . . I believe the officers of the Army will be in demand in air transportation companies."[70] Patrick advised that the young man's future should rest with the young man: "his own inclinations should govern largely."

If there was one issue that all members of the Air Service agreed on, it was civil aviation development. Hap Arnold started publishing articles about it in 1920.[71] Patrick's articles and speeches concerning civil aviation increased dramatically in the years 1924 and 1925. In each case, he would invariably tie progress in civil aviation to the nation's proper defense: "we believe that the development of [civil] air transportation has a decided bearing upon the solution of the problem of National Defense."[72] Patrick would then follow up with a plea for appropriate federal regulations that would ensure the safe, rapid development of a federal airway system: "Of all the important nations of the world, the United States is the only one that has not enacted suitable legislation to regulate flying and to safeguard the interests of the people."[73]

Patrick made this observation in 1925. Back in 1921, Herbert Hoover, who was then Harding's secretary of commerce, was trying to deal with the issue as well when he wrote, "It is interesting to note, that this is the only industry that favors having itself regulated by Government."[74] The real problem regarding regulating legislation for civil aviation lay in the fact that Congress really had no mandate to act on the issue; a marginalized constituency wanted this done. And its advocates wanted regulation, not for monetary reasons or because of unfair competition, but primarily for safety's sake. Civil aviation was simply not an emotional grassroots issue with Congress.[75] It did not get votes.

Patrick, in a January 1925 article entitled "Air Service and Air Transportation," wove together the components of his civil air transportation program: interest, infrastructure, and federal regulation.[76] Patrick began to push the idea of civil air transportation in earnest in October 1924 following the successful conclusion of the Air Services' "Around the World Flight." This project, which took place over 175 days in 1924, was Patrick's plan from the start. Though many entertained the idea, Patrick took the possibility seriously enough to attempt it. The significance of this particular project is attested to by his interest in all facets of this undertaking and the prominent role he played in publicizing it. Patrick ordered his staff to do preliminary work on the project in early 1922, taking a personal hand in all aspects of the endeavor.[77] The U.S. State Department, the U.S. Navy, Coast Guard, the Geological Survey, and even the Bureau of Fisheries were involved in the great adventure.[78] Four specially designed and built Douglas World Cruisers left Seattle on 6 April 1924. The route went by way of Alaska, Japan, India, Turkey, England, and Greenland, with many stops in between. With the loss of only one plane (the crew survived) three of the original planes and crews returned to Seattle on 28 September.[79] Patrick had orchestrated a grand public affairs bonanza, and it fit in perfectly with his effort to publicize civil aviation.

Patrick used the dramatic success of the World Flight as proof of aircraft's capability to traverse long distances dependably and safely. With each dramatic success in the air, whether it was the pre-Patrick era transcontinental race of 1919, a cross-country flight by Lt. Jimmy Doolittle, the first in-flight refueling tests, or attempted altitude records with turbo-charged engines and oxygen-equipped aircraft, Patrick was able to point with pride at the groundbreaking efforts of his Air Service. Each accomplishment was designed in some fashion to demonstrate aircraft reliability over ever increasing distances. The Air Service, was, indeed, leading the way in publicizing the promising future of air transportation for the general public.

This aspect of Patrick's overall public affairs program for civil aviation is nowhere more evident than in the number of appearances he made before insurance and underwriting audiences. Patrick gladly addressed the annual meetings of the Association of Life Insurance Presidents and other related organizations.[80] The newsletter of one such insurance association, in announcing Patrick and Illinois Central Railroad president Charles Markham as the featured speakers at an upcoming convention, described them as "[t]he heads of two great American transportation systems."[81] This, of course, was in reference to Patrick's management and ongoing organization of the over 4,000 landing fields in the United States to aid in the development of commercial aviation. By calming the underwriting fears of the insurance companies, Patrick was securing one of the major financial building blocks required to further expand the nation's fledgling commercial aviation sector.

Ultimately, Patrick's personal efforts paid off when, in February 1925, Congress passed the Air Mail Act (or Kelly Bill), which directed that the Air Mail Service be assigned to private companies. This effort predated commercial passenger transportation and, in effect, set the stage for that endeavor.[82] The federal air-mail contracts provided the subsidy that enabled commercial airline activity to get established on a firm footing. Indeed, by 1928 two regularly scheduled airlines cooperated in a coast-to-coast route.[83] Patrick pushed vigorously for a national policy to govern the aviation industry, which was part of his responsibilities as outlined by the War Department.

Many of Patrick's goals were also shared by the National Aeronautic Association (NAA) and the National Advisory Committee for Aeronautics (NACA). Patrick was on the job for less than a week when Dr. Carter Ames, a key member of the NACA, visited Patrick. The topic: legislation.[84] This issue was not new to Patrick, having testified for federal regulation of aviation two years prior.[85] Patrick also did not get any help from Billy Mitchell on this issue as Mitchell wanted the Air Service to control all civilian aviation.[86] Civilians of all stripes rejected this idea outright. It fell to Patrick to come up with a compromise to resolve the issue. He worked closely with the NAA and the American Legion and pushed for appropriate legislation between 1922 and 1925.[87] Although he was unsuccessful in getting anything through Congress during those years, the tide was beginning to turn.

By this time, congressional thinking had changed concerning civil aviation, and

Congress considered rectifying the many Air Service deficiencies. Patrick did not have tunnel vision when it came to working the military and civil aviation programs. He was well aware of the symbiotic relationship between the two communities. The healthier each entity, the better for both. In addition, both were supported by one and the same aviation manufacturing base. Patrick's overall effort to support all three legs of the aviation triangle was part of his long-term vision to make the United States a true air-faring nation.

Patrick's personal interest and emphasis regarding the commercial aviation industry is little known or appreciated. He deserves much of the credit for trailblazing the airway system that was ultimately used by every commercial airline enterprise during the 1920s and '30s. He also deserves recognition for making Congress, the business community, state and city governments, and the public at large cognizant of the importance and future promise of commercial air transport.

10. The Fallout from the Lassiter Report and the Fall of Billy Mitchell

After the Lassiter Board

From the time of the Lassiter Board in April 1923 through 1924, Patrick had an exceptionally busy schedule.[1] He was involved in efforts aimed at correcting legislation concerning both civil and military aviation matters. If he was not in Washington, he was on the lecture and speech circuit drumming up support for his program. Patrick kept the Air Service in the public's eye and in the eye of Congress in other ways as well. Doubtless, the most newsworthy event was the Around the World Flight. But there were also attempts at transcontinental records, national and internationally sanctioned air races, and the testing of new aircraft and equipment. Patrick was squarely behind pioneering efforts to employ the unique talents of aircraft: cropdusting for the Department of Agriculture; the resumption of fire patrols in the West Coast states; and geographic surveying and aerial photography.[2] Many of these initiatives were the work of Patrick's dedicated headquarters staff and officers out in the field; for example, as the commanding officer at Rockwell Field, Hap Arnold was responsible for instituting the valuable fire patrols.

Overall, the work Patrick accomplished in 1923–24 can best be described as a calculated offensive strategy that emphasized the cultivation of support from private, professional, and public forums. Where Mitchell was dogmatic, with calculated criticisms directed at his enemies, especially the General Staff and the Navy, Patrick was accommodating, yet persistent. Patrick was successful in controlling Mitchell during this period, largely because he sent him either overseas (thirteen months) or on various inspection tours in the states.

A perceptible shift occurred in how the American media viewed the Air Service and Patrick's efforts. During 1923–24, for example, the *New York Times* printed forty-two news features relating to Patrick. Without exception, the seven editorials in which Patrick was featured within that same time period were all favorable. By comparison, there were only sixteen news articles relating to Billy Mitchell (including three about his engagement and marriage to Elizabeth Miller) and only one editorial that mentioned him in passing.[3] Almost from the moment that Patrick offered to accept his resignation in October 1921, Mitchell said or printed very little that was controversial except for his comments following the bombing of the *Virginia* and *New Jersey* in September 1923.

Others factors were limiting Mitchell's freedom of expression as well. In October 1921 Harbord and Patrick prohibited the publication of articles without their approval. Secretary of War Weeks forbade Mitchell to publish anything without the express approval of the War Department. This particular prohibition was issued shortly after Mitchell's *Ostfriesland* bombing report was leaked to the press.[4] The combined admonitions had the desired effect. Indeed, out of the total of Mitchell's writings (three books and 108 articles), for all of 1923 and 1924 only five articles were published.[5]

Patrick was actually quite satisfied with Mitchell's performance during this period. In fact, his comments in Mitchell's 1923 officer efficiency report reflect that Mitchell had stayed within bounds as Patrick had proscribed, but the Air Service chief did have some reservations: "He has shown considerable improvement, but it is difficult for him to subordinate his own views and opinions to those of others."[6] Capt. Ira Eaker, Patrick's assistant executive officer, was in a unique position to view the relationship: his office was right between Patrick's and Mitchell's. Eaker, both brilliant and tactful, encapsulated perfectly the view Patrick had of Mitchell:

Patrick admired Mitchell's ability but didn't approve his stepping outside military procedures to accomplish ends. Patrick spoke to [Mitchell] severely several times about the methods he was supposed to use. It was a question of evolution versus revolution, these separated Patrick and Mitchell. To a certain extent, Patrick was right. He got more for the Air Corps than anybody else could have gotten.[7]

But Patrick himself was chafing at the inaction of the War Department and Congress to address the needs of the Air Service. From April 1923, when Secretary Weeks approved the Lassiter Board findings, until the end of 1924, Patrick maintained a fervent educational effort that was twofold. One effort was aimed at the Army leadership; the other was a methodical crusade to bring the knowledge of aviation capability to the public and to private enterprise. From his first day as Air

Service chief, Patrick proceeded to assuage the oft-times antagonistic feelings the Army leadership held toward the Air Service. The success of the Lassiter report is evidence that he was very successful in this regard, even winning over the likes of Brig. Gen. Hugh Drum, who could always be counted on to be critical of the Air Service.

Early in Patrick's tenure there is clear evidence that the senior Army leadership was sensitive to the needs of the air arm. About two months after Patrick took over, Patrick had his staff (with the help of Capt. William Sherman at the Tactical School) draft an article about the Air Service and military aeronautics. Patrick sent this article forward for Pershing's signature and subsequent publication in the *Aeronautical Digest.* Patrick had written, in part, that "an Army of the future must possess an up-to-date, adequate, efficient, highly trained Air Force." Pershing's staff, and in particular, Col. George C. Marshall, added the clarification that "within the limits of appropriations the War Department is doing everything in its power to ensure that end."[8]

The Air Service, according to the War Department, was an expensive branch of the Army. But as a percentage of the overall War Department appropriation, the Air Service budget was significantly reduced each year after the war until 1924. When Patrick took over the Air Service, he inherited a budget that was 11.2 percent of the total War Department budget. In 1922 the Air Service budget was cut in half and then halved again in 1923, resulting in an Air Service budget of $12.9 million or only 3 percent of the War Department budget for 1923. Due to Patrick's efforts, however, the percentage rose incrementally the next year and significantly each year thereafter with the largest increases taking place when the five-year Air Corps Act appropriations took effect (see Table 1).[9] In all fairness to the Army, drastic cuts came not only to the Air Service but to all branches.[10]

Budgets were indeed tight for all branches of the Army, and there was a marked disdain among ground officers for the very public battle the Air Service waged for increased funding. Patrick received a very personal example of those feelings when he wrote to a friend of long standing, Maj. Gen. Fred W. Sladen, the superintendent of West Point, in February 1925, asking if possible, that the nonfunctioning "Patrick Fountain" be "re-instated." Sladen responded: "[I]t will cost some $300 to re-instate the Patrick Fountain. You and the Air Service may not feel the pinch of poverty. We here do, and I do not feel that I can spare the price to restore the fountain."

The General Staff was concerned because aircraft acquisition was extremely expensive. If the Army had actually implemented "War Department Major Project Number 4" (the Lassiter Program) without any increase in the War Department budget, the Air Service would have eaten up almost 12 percent of the Army budget.[11] Brig. Gen. Fox Conner, the G-4 on the General Staff, had directed a

Table 1
War Department and Air Service/Air Corps Appropriations, 1921–1930

Year	War Department	Air Service/Air Corps[a]
1921	297,320,000	33,435,000
1922	388,248,000	25,298,000
1923	418,700,000	12,895,000
1924	348,524,000	12,676,000
1925	349,401,000	13,927,000
1926	293,913,000	16,850,000
1927	302,209,000	19,050,000
1928	332,212,000	22,191,000
1929	355,123,000	26,435,000
1930	398,364,000	34,690,000

Note: War Department appropriations compiled from annual reports of the secretary of war for the respective years 1921 through 1930; 1921 through 1925 Air Service appropriations compiled from annual reports of the secretary of war; 1926 through 1930 Air Corps appropriations from Chase C. Mooney, *Legislation Relating to the Army Air Forces Materiel Programs, 1939–1944,* Army Air Forces Historical Study no. 22 (rev.), Aug. 1949, appendix 1, p. 162.

[a]Air Service, 1919–26; Air Corps, 1927–30

1925 study of the financial implications of the Lassiter program. He noted, quite rightly, that without additional funding by Congress the War Department could not even begin to implement the Lassiter requirements. "There would be little left for the rest of the Regular Army."[12] Conner's remark was hyperbole but, realistically, Lassiter could not be funded out of hide.[13]

This funding limitation, combined with the theoretical arguments surrounding the capabilities of the aviation weapon system, created an emotional mix. Patrick understood this environment. By comparison, it would be easy to say that Mitchell did not, which would provide an understandable rationale for Mitchell's bombastic approach to the issue. But that justification does not hold up. Mitchell understood the budgetary limitations the War Department was working under. This was all the more reason for him to seek a separate "Ministry of Defense" that would contain an independent Air Force Department co-equal with the Army and Navy.

Patrick successfully contained Mitchell's rebellious spirit through 1924, but Mitchell's emotions were about to burst forth again, despite Patrick's best efforts to contain them. Mitchell entered the headlines again as a result of the Lassiter Board plan.

As 1924 came to a close, Patrick was indeed frustrated, probably as much as Mitchell, that little had been done by the War Department to redress the shortcomings of the Air Service as outlined in the recommendations of the Lassiter Board. Due to Patrick's lobbying efforts with his acquaintances on the General

Staff, and the good relationship he developed with Congress, Patrick was now convinced that these people understood the need for change in the Air Service.[14] The General Staff's acceptance of the Lassiter recommendations was clear evidence of this. While the General Staff still believed in the primacy of the ground forces, they acknowledged three key issues: the role of an independent air force mission removed from direct ground support; the requirement to bolster the tactical Air Service; and the necessity to have an adequate Air Service in place at the onset of hostilities.

Patrick had successfully converted the majority of the General Staff to his way of thinking on several key issues. If there were any doctrinal differences, they were not contentious, for they were being resolved with the ongoing coordination of Training Regulation 440-15. It was a problem of funding, plain and simple. President Coolidge's economy government was the real culprit. No less a personage than the Army Chief of Staff Gen. John J. Pershing concurred. Prior to his retirement as chief of staff in September 1924, Pershing sent his final report to Secretary Weeks. Pershing stated emphatically that the Army's funding must be increased to support critical expansion in the Army across the board, even at the expense of overseas garrisons.[15]

In all fairness, the War Department did what it could to obtain funding to implement the Lassiter recommendations, especially concerning the expansion of the peacetime Air Service. It forwarded those recommendations to the Joint Army and Navy Board for action. This group of individuals, headed by the Army chief of staff and the chief of naval operations, would hammer out any differences in proposed programs prior to being forwarded for congressional action. The Lassiter recommendations were forwarded to the Joint Board shortly after the War Department approved them, and there they lingered. The secretary of war and the secretary of the Navy could not agree on the division of the recommended appropriation. The Lassiter program, as approved by the War Department, directed the Army and Navy to join together in requesting aviation appropriations from Congress. The Navy had a five-year aviation procurement program in the works at the time. Secretary Weeks proposed that the Army and Navy enter into a cooperative ten-year acquisition program. This would reduce the annual cost, thus making it more acceptable to Congress. But Weeks tacked on the requirement that the appropriations, if and when available, be split 60/40, with the larger share going to the Army. Navy Secretary Edwin Denby and his successor, Curtis Wilbur, said no. There the Lassiter program sat, at an impasse.[16]

The Lassiter program stalled out in the Joint Board in February 1924.[17] Secretary Denby diplomatically informed Secretary Weeks that the U.S. Navy would gladly assist the War Department in its quest to improve the aviation program, but the Navy was not going to sacrifice its appropriations to do so.

The Lampert Committee

Coincidentally, in October 1924, the Lampert Committee began examining charges of undue influence by the Manufacturers Aircraft Association on the Air Service. The committee had an extraordinarily wide charter to investigate anything "in any way connected with any and all transactions of the said United States Army Air Service," and it quickly expanded its probe to examine military aviation in general.[18] The deliberations of this committee turned out to be the catalyst for breaking loose the Lassiter program that was long deadlocked in the Joint Board. This, along with other actions, ultimately resulted in legislation that led to the Air Corps Act of 1926. The Lampert Committee also turned out to be the forum where Billy Mitchell, after two years of relative quiet, finally "broke out," as Patrick described it, and was fast on his way to becoming the very public martyr that his actions destined him to be.[19]

Mitchell's time was running out. His four-year term as assistant chief of the Air Service was due to expire on 26 April 1925. Mitchell, having returned from his extended honeymoon/inspection tour of the Pacific and Far East in July 1924, arrived back at Air Service headquarters when Patrick and his staff were preparing for testimony before the Lampert Committee. By then, the committee members were examining all aspects of Air Service funding, organization, and equipment, as well as the issue of independence.

In a 19 December letter to the adjutant general, Patrick suggested the establishment of an Air Corps as an intermediate step prior to the creation of an independent air force within a national defense organization.[20]

I am convinced that the ultimate solution of the air defense problem of this country is a united air force, that is the placing of all the component air units, and possibly all aeronautical development under one responsible and directing head. Until the time when such a radical reorganization can be effected, certain preliminary steps may well be taken, all with the ultimate end in view.[21]

The significant "preliminary step" that Patrick recommended was the creation of an Air Corps analogous to the relationship between the Marines and the Navy. This proposal created quite a stir in the War Department. Forwarded to the War Plans Division for evaluation, the Air Corps concept was promptly disapproved due to a concern with "unity of command." But they agreed with Patrick on the need for more funding. General Drum, though, wanted to challenge Patrick's proposal: "If his contentions are correct, remedies should be applied; if they are not correct, they should be withdrawn." Drum was overruled, and Patrick's proposal was not withdrawn.[22]

Patrick's letter suggesting the creation of an Air Corps as an intermediate stage prior to independence set the stage for his testimony before the Lampert Committee. One of over 150 witnesses that came before the committee, he testified on 5 January 1925. Ultimately, in the six volumes of testimony taken by the committee, Patrick provided the only novel approach to the contentious issue of independence. In addition, he used the Lampert forum to push forcibly for the Air Service to alone be tasked with the coastal defense mission. Both of these ideas were well received by the committee members. Although the Lampert Committee may have expressed an interest in the Air Corps concept, the War Department response was nil.

Mitchell's Fall

Having returned from his Pacific inspection tour in July 1924, Mitchell was obviously reinvigorated concerning the promise of air power. His 323-page "Report of Inspection," which he forwarded to General Patrick, was a blueprint of the next war in the Pacific theater as Mitchell envisioned it. The report was also highly critical of Maj. Gen. Charles P. Summerall's administration of the Hawaiian Department.[23] When Mitchell returned home, "[h]e commenced a vigorous campaign to win acceptance for his ideas on air power." His patience at an end, he "wanted his program approved completely and immediately." [24] Patrick noticed this new stridency in his deputy's demeanor and noted in Mitchell's 1924 efficiency report that "[h]is recommendations frequently fail to take into account conditions actually existing and which must be, in a measure, controlling. . . . His opinions on many matters are frequently biased."[25]

As Mitchell's tone became more strident in his speaking engagements, his penchant for agitation began in earnest when Coolidge sent Mitchell as his representative to speak to a convention of the National Aeronautical Association in October. Mitchell addressed some sensitive issues: bombing of civilian targets, the possible decisiveness of air power in future wars, and how surface navies were obsolete.[26] He also agreed to publish a series of articles in the *Saturday Evening Post*. Not only were these articles inflammatory, but the way in which Mitchell went about getting permission to publish was open to question. Knowing that he needed permission, he bypassed Secretary Weeks and went directly to President Coolidge, who gave tacit permission if Patrick approved. Patrick acquiesced, but Mitchell did not tell Patrick of a follow-up letter from Coolidge directing Mitchell to abide by the direction given him by Patrick and Weeks. Mitchell bypassed Weeks completely. Mitchell had his series written and in print without anyone in his chain of command having reviewed them.[27] Weeks was not pleased.

During his repeated appearances before the Lampert Committee, Mitchell's pent-up frustration burst forth in a litany of accusations against the War and Navy Departments regarding the treatment of the Air Service.[28] His criticisms were many and varied. He was forceful in his urgings for a department of national defense.[29] He accused the General Staff of stifling honest testimony of junior officers; he saw a cabal of vested interests, including the Army, Navy, the Air Mail section of the Post Office, and the NACA, of conspiring to deny the full development of the nation's air power; and he demanded an independent air force. Mitchell only rubbed more salt into not-yet-healed wounds when he stated, "It is a very serious question whether air power is auxiliary to the Army and the Navy, or whether armies and navies are not actually auxiliary to air power."[30]

Secretary Weeks was incensed by Mitchell's tirade and demanded that he prove his charges. Mitchell's detailed and patronizing justification for his testimony ended any chance of his continuing as assistant chief of the Air Service. Weeks asked General Patrick to recommend another officer to take Mitchell's place, even though Patrick had recommended that Mitchell be reappointed.[31] The culmination of Mitchell's wayward independence had taken its toll on Secretary Weeks's patience. Patrick selected Lt. Col. James Fechet, then commanding the Advanced Flying School in San Antonio, to replace Mitchell, effective 26 April 1925.[32]

The only problem that seemingly remained for Patrick was where to station Mitchell. Patrick discussed this issue with several of his friends. One in particular, Maj. Gen. George B. Duncan, Commanding General of the Seventh Corps Area at Omaha, Nebraska, campaigned to get Mitchell as his commanding aviation officer. Duncan gave voice to what other corps commanders were doubtless thinking in an 11 March 1925 letter to Patrick:

I see that you are going to get rid of your stormy petrel of the Air Service away from duty in Washington. It occurs to me that with the controversy it would probably be your intention to keep him off a coast assignment . . .

. . . I could use Mitchell to good effect . . .

If you decide to send him here I will be glad if you could write out for me instructions that he must follow in the matter of public speaking, that he would know at once what his limitations would be and what I expected of him, and any departure from this would be followed by disciplinary action.[33]

Patrick, in his reply to Duncan, refers to Mitchell as "my soon to be Ex-Assistant." With purposeful understatement, he noted: "He does have his good points, and . . . he can stimulate interest. Just where he will go is not yet settled . . . although I know that you would do everything possible to keep him under control."[34] Shortly thereafter, Mitchell reverted to his permanent rank of colonel and took his position as air officer of the Eighth Corps Area at Fort Sam Houston in San Antonio, Texas.

Mitchell's departure was amicable. He had no hard feelings toward Patrick, as demonstrated during his farewell luncheon on 27 April, which Patrick hosted. A *National Aeronautic Association Review* article noted how General Patrick, as the event's toastmaster, described Mitchell: "In introducing the departing officer, General Patrick paid him a high tribute, saying that he believed General Mitchell had done more for the Air Service than, perhaps, any other individual man."[35]

As far as Mitchell's demotion and banishment are concerned, there is little firsthand documentation concerning why he decided to speak out when he did. Mitchell does, though, allude to his exasperation in a letter to Hap Arnold prior to the Pacific inspection tour, noting that "air power doesn't seem to be getting anywhere at all. The public's interested, but people in Washington who could do something about it aren't."[36] Upon Mitchell's return to the States in the summer of 1924, he saw that little more had been done. In fact, Congress was reducing Air Service appropriations and the department was further reducing manning and equipment of the already anemic Air Service.[37] Mitchell saw little hope that current initiatives would yield anything resembling his quest for the realization of air power as he envisioned it.

Mitchell had been sidelined and muzzled for almost three years, ever since the *Ostfriesland* bombing. During that time he was detailed overseas on two lengthy inspection tours: to Europe and the Pacific. While Mitchell returned from his Asian tour fully energized by what he saw as the necessity to prepare for the next war, he found that the War Department did not share his sense of urgency, and that Patrick had not made any progress either. Presented with the opportunity to speak his mind before the Lampert Committee, Mitchell did so, with a vengeance. What is interesting is that given Mitchell's outspoken behavior, Patrick still defended him and recommended his reappointment as his deputy. This is clear evidence that Patrick and Mitchell were philosophical soulmates regarding airpower doctrine, but certainly disagreed about the way to implement such ambitious dreams. Patrick used Mitchell to push the envelope of public and congressional opinion. Mitchell, an airpower evangelist who was banned from the pulpit for past sins against the hierarchy, loudly announced his beliefs to all who would listen. He was banished to the hinterland of the Air Service realm not only for what he said and wrote, but also for the stridency of his tone and his unorthodox public approach.

While Mitchell may have been exiled, he did not remain quiet for long. In the autumn of 1925, two naval air accidents precipitated Mitchell's court-martial. The first incident involved an attempt by three Navy planes to reach Hawaii from San Francisco. They departed on 31 August; two planes were lost early on and the crews recovered. When the third plane went down, the crew was feared lost but ultimately recovered. While the search continued for the third aircraft, the Navy's huge dirigible *Shenandoah* went down in a squall while crossing over Ohio. Cdr.

Zachary Lansdowne, a friend of Mitchell's, perished along with twelve other men. Reporters began calling Mitchell for a statement about the tragedy. In an ensuing press conference on 5 September, Mitchell read a prepared statement: "These accidents are the result of the incompetency, the criminal negligence, and the almost treasonable negligence of our national defense by the Navy and War Departments."[38]

President Coolidge himself signed the court-martial orders against Mitchell. In addition, Coolidge followed through on a proposal he had originally suggested to his Amherst college classmate, banker Dwight W. Morrow, several months earlier. On 12 September 1925 Coolidge appointed Morrow to chair a board to study "the best means of developing and applying aircraft in national defense.[39] The Morrow Board, otherwise known as the President's Aircraft Board, was tasked to investigate the same issues that the Lampert Committee had investigated over a five-month period (October 1924–March 1925). At the start of the Morrow Board on 17 September, the Lampert Committee had still not issued its report. Morrow and company had no such desire to drag out their investigation.[40] First, Coolidge wanted to preempt the issuance of the long-awaited Lampert Report (which would be pro–Air Service independence) with a much more conservative recommendation, and, second, Coolidge did not need to fund the Morrow Board. As the president explained, the committee members were "all sufficiently prosperous not to be handicapped seriously by the question of finances."[41] Before the Morrow Board issued its report on 30 November 1925, two weeks prior to the issuance of the Lampert Report, the board took testimony from ninety-nine witnesses. Among them were Patrick, Mitchell, Foulois, and Majors Arnold, Walter G. Kilner, Thomas DeWitt Milling, and Horace M. Hickam. Patrick's testimony to the Morrow Board was a repetition of his words to the Lampert Committee: the Air Service is in dire need of increased funding, new aircraft, a separate promotion list, and the need to create an Air Corps as a transitional organization prior to the realization of an independent air force at some future time.[42] Of note is the fact that nearly all Air Service officers who testified supported Patrick's testimony. The notable holdout was Billy Mitchell. As in the case of the Lampert Committee, Mitchell's testimony before the Morrow Board was at odds with Patrick's and the majority of his Air Service comrades.[43] Mitchell was in the small minority urging immediate independence for the Air Service within a national defense organization.[44]

While many junior officers of the Air Service rallied behind Mitchell during this period, the more senior officers (and ex–Air Service officers) felt much differently. Edgar Gorrell was quite critical of Mitchell's tactics and complained to Patrick in writing a few days after Mitchell's 5 September San Antonio press conference.

I read Colonel Mitchell's recent publicity with great chagrin. Needless to say that many things he said in his publicity are not correct.

I want you to know the general opinion I find among the public is that Mitchell is not trying to help the United States, but is trying to feather his own nest and further his own personal prestige.

The thinking public believe[s] the War Department should take a definite stand and tell the public the truth about Mitchell. He has, of course, captured the minds of the unthinking public, but if the War department will come out and tell the truth about him, the time is psychological to shut him up.[45]

Patrick's reply was tinged with a note of exasperation.

About Colonel Mitchell's recent pronouncement, I believe your letter reflects the view of many thinking people. For many reasons I am sorry that Mitchell saw fit to give out such a statement. It was most intemperate and in many ways unjust. I think it will merely befog the entire situation. I had hoped for Congressional action which would improve conditions. So many things have been interjected now that the prospect is not bright. I fancy we shall have a hectic winter with more Congressional investigations, added work and annoyance.[46]

Patrick was correct. The prospects were not bright. The Lampert Report was being written for release, the Morrow Board had witnesses standing in line to testify, and Billy Mitchell was about to go on trial.

Coolidge convened the court on 20 October, and the trial began in Washington on 28 October 1925. Mitchell was charged under the 96th Article of War: "all disorders and neglects to the prejudice of good order and military discipline, all conduct of a nature to bring discredit upon the military service." It only had to be proven that Mitchell had brought discredit upon the service in very broad terms for him to be found guilty. Mitchell entered a plea of not guilty ("very loudly," according to reporters) and succeeded, along with his civilian defense counsel, in drawing the trial out for seven weeks. The trial was turned into a referendum on the issue of military aviation and how the national defense was managed. Found guilty as charged, Mitchell was sentenced to suspension and forfeiture of pay and allowances for five years. Coolidge approved the sentence on 26 January 1926 but reduced the sentence to full subsistence and half pay. Mitchell offered his resignation effective 1 February 1926, and the War Department accepted it.

Patrick was one of the witnesses for the prosecution. His testimony centered on the types and condition of aircraft flown by the Air Service, followed by a cogent defense of the Air Service as coping as best it could, given the limited funding. Patrick specifically attributed the lack of Air Service funding to the Bureau of the

Budget. While presenting a picture that pretty much mirrored Mitchell's criticisms, he pointed out that Mitchell himself was part of the decision-making process.[47] Patrick walked both sides of the issue. He could not very well deny all that Mitchell said without seriously undermining his own testimony given before numerous investigating committees. Patrick walked a fine line during the trial and did so successfully. In later years Ira Eaker pointed out that Patrick's role as a prosecution witness may have "cost him the credit deserved for his role in fostering the early development of American air power."[48]

Patrick had to walk a fine line outside the courtroom as well. He said nothing in public about the trial and, according to Eaker, strongly cautioned the officers on the Air Service staff to do nothing that would jeopardize their careers.[49]

If Mitchell's court-martial was responsible for one thing, it may have contributed to the hurried release of the Morrow Report.[50] It was released just as Mitchell's courtroom defense came to an end, bumping the trial off the front pages. But there was a more important result, for Coolidge's calculated effort paid off. It greatly overshadowed the soon-to-be-released Lampert Report. While the Lampert Report would recommend Air Service independence, the Morrow Board called for an Air Corps, along with an assistant secretary for air, and additional personnel and a five-year funding program. This was a proposal that the War Department might accept.

11. The Air Corps Act and Its Aftermath

The Time to Act Is Now
With the culmination of Billy Mitchell's trial, combined with the publication of the Morrow Board and Lampert Committee reports, an opportunity had come for the introduction and passage of new aviation legislation. Patrick need not have worried about new investigations. He was well aware of this unique juxtaposition of activity and felt resolution in the air: "I believe the work we are doing is bearing fruit, that there is more interest taken in the development of aeronautics now than ever before in this country and that something will come of it."[1]

All interested parties to this great drama had heightened expectations. It was not so much a question of *whether* there would be Air Service legislation as *when*, and in what form, it would occur. The public was aware of the headline-making news regarding the Air Service and wanted to see some reform. The aviation manufacturing industry, having been investigated at length, along with the Air Service, lobbied for change. The General Staff, not to mention the Army in general, had also received an extended education on the need for a new view of military aviation. Patrick and Mitchell had prepared this ground exceedingly well, at times contentiously on Mitchell's part, but balanced by Patrick's well-reasoned arguments. Congress itself, having instigated and completed its investigations, overflowed with legislative proposals.

However, the most important factor for impending change was none of these parties. The critical spark was President Coolidge. He set the Morrow Board in motion, and the results of that board offered to the president an expedient way of getting the aviation and Air Service problem off the table. Coolidge took a more direct approach with Mitchell and rightfully so. But the president had to contend with the other legislative proposals in the works.

December 1925 witnessed two bills hurriedly introduced by various members of the 69th Congress, even before Mitchell's trial came to a close, doubtless as a sign of support for the besieged airman. These efforts contained nothing new, lacked the beneficial weight of the Lampert Committee and Morrow Board reports, and did not even get out of committee. But within a week after the Lampert report was out, Rep. Charles F. Curry of California, a long-time advocate of a separate air force, introduced H.R. 447, the first bill to recommend a department of national defense with three coequal departments: land, sea, and air.[2] Soon thereafter, two similar House proposals were thrown into the legislative hopper.[3] It was not until 18 January 1926 that the War Department finally entered the fray as well, knowing full well that their legislative proposal, the Air Bill (H.R. 7916 by Rep. John Morin of Pennsylvania), would have the backing of the president as it was based on the Morrow Board results.[4] The Morrow Board, of course, opposed both a national defense scheme and an independent air arm. It recommended an Air Corps organization that would be represented on the General Staff, as well as a dedicated assistant secretary for air.

Thus it was that on 19 January that the House Committee on Military Affairs met to sort out the best approach to the military aviation problem. The options boiled down to three: a national defense organization with three equal agencies for the Army, Navy, and Air Force; retaining the current structure but simply adding an Air Department; and last, making the Air Service an Air Corps within the Army.

The Air Corps option was attractive to many on the committee, so much so that the House Committee wrote to Patrick asking specifically that he expand upon the idea which he first proposed to the adjutant general on 19 December 1924.[5] In that original memorandum to the adjutant general, Patrick even included "A Statement Containing a Summary of the Organization, Development, History, and Mission of the Marine Corps," written by the Marine Corps commandant, Maj. Gen. John A. Lejeune, on 28 January 1924. A prominent part of that proposal was Patrick's conviction that "the ultimate solution of the air defense problem of this country is a united air force."[6] Patrick reiterated this belief when he forwarded his expanded answer to the committee. Patrick's original five-page Air Corps memorandum may have been ignored by the War Department but not by the House Committee on Military Affairs. In reality, this proposal, buried by the War Department, had two resurrections, initially during Patrick's testimony before the Lampert Committee and subsequently before the Morrow Board.[7] The War Department did not query Patrick about his December 1924 Air Corps initiative when he explained it during his Lampert Committee testimony, but when he mentioned it during his Morrow Board testimony, the War Department hurriedly tasked Patrick to complete and forward a detailed commentary "in about four working days."[8]

Whether Patrick spoke before the Lampert Committee or the Morrow Board, he always explained that he envisioned his Air Corps proposal as nothing more than an interim arrangement prior to the ultimate creation of an independent air force within a Department of Defense.[9] But the political climate would not allow it just yet. Thus, when the House Committee on Military Affairs asked Patrick to draft a bill that centered around the Air Corps concept, Patrick obliged. He forwarded this draft to the Military Affairs Committee through the secretary of war, and what came to be termed the "Patrick Bill" was ultimately incorporated into H.R. 8533, which was originally introduced by Rep. J. M. Wainwright on 28 January.[10] What the Wainwright Bill attempted to accomplish was a compromise between the advocates of the Morrow Board approach and the Lampert approach. The Morrow Board found expression in the Morin Bill (H.R. 7916), drafted by the War Department, which recommended:

1. The name of the Air Service be changed to Air Corps;
2. An assistant secretary of war for air be established;
3. Each General Staff division have an Air Corps representative; and
4. A five-year aviation procurement program be authorized (a little over half the amount as recommended by the Lassiter Board).[11]

The Lampert Committee's recommendations centered around the establishment of a Department of National Defense and the proposals that Patrick had made to the Lassiter Board. These issues were put forth in the form of the James Bill (H.R. 9044) introduced on 8 February 1926 by Rep. W. Frank James of Michigan.

The Wainwright Bill proposed a separate budget for the Air Corps, dedicated maintenance as well as training facilities, and the control of all aerial operations from land bases. This would be accomplished under the direction of a new assistant secretary of war for air and, thus, the Air Corps would be removed from direct Army control. Given the autonomy the Air Corps would receive under the provisions of the Wainwright Bill, it was anything but a compromise and was harshly criticized by both the Army and Navy Departments.

On one side of the legislative fence stood the Executive and War Departments with the Morin Bill, which offered minor changes in the Army/Air Service relationship. On the other side of the fence stood two legislative proposals: the Patrick/Wainwright combination and the even more drastic James Bill. Patrick hoped that a combination of the Patrick/Wainwright and James Bills would have yielded the best of all legislative worlds.[12] But from where Patrick sat, he knew that there would be Air Service legislation passed by this Congress. And it would include, to some extent, proposals that he himself had initiated, even though those proposals did not include all that he thought necessary.

For nearly three years the War Department knew how Patrick felt about the expansion of the Air Service, and precisely how he would go about it with his proposed five-year plan. The plan was studied and endorsed by the Lassiter Board and the secretary of war and then died of inaction in the Joint Board. Patrick labored long and hard to put this proposal in the legislative forums of the House and Senate. Now that action was going to be taken, Patrick was aware that the Lassiter program, as exemplified by the James Bill, was in competition with the Patrick/Wainwright legislation that he had a direct hand in formulating. But, if anything, Patrick was pragmatic. After three years, he was willing to settle for half a legislative loaf, with the hope of catching up later.

What that half a loaf could include became clear when the War Department in late December 1925 asked Patrick to draft an Air Service expansion program based on the five-year acquisition budget proposed by the Morrow Board. The "half a loaf" victory was literally true. The Morrow Board would spend no more than a little over half of what Patrick had recommended to the Lassiter Board. On 8 January Patrick responded to the adjutant general's request with a detailed expansion within the guidelines of the Morrow Board. Patrick did so only after voicing his objections:

I consider it a mistake for the War Department to depart from [Lassiter] policy as it is the ultimate expansion program for the Air Service and I do not concur in any less ideal solution than that proposed by the Lassiter Board and approved by the Secretary of War. It is noted that since the approval of the Lassiter Board there has never been any doubt as to the need for the war organization prescribed.

After noting his reservations, Patrick could only hope that the shortfall would be redressed at some future point.

However, in conformity with the recommendation of the President's Aircraft Board [Morrow Board], I am of the opinion that for the next five years the War department might well have a program for the expansion of the Air Service along the lines of and somewhat less than that proposed by the Lassiter Board, so that at the end of five years a balanced Air service organization will exist, and if it is deemed expedient at that time the remainder of the Lassiter Board organization can constitute a subsequent program for the further expansion of the Air Service.[13]

While the War Department and the House were shaping their legislative compromises on an Air Service bill, the Senate was doing the same thing almost in mirror-image fashion. Sen. William Hill's legislative proposal was much along the lines of the Lampert Committee with regard to a Department of National Defense. Its opposite number was a bill by Sen. James Wadsworth which mirrored the Morrow recommendations.[14] The House Committee on Military Affairs asked for testimony from many familiar faces who had made the rounds of the Lampert Committee and the Morrow Board. The same old words were spoken. Nothing new was said. But there was a sense of urgency because everyone knew that some type of legislation would be passed: the president, House, Senate, War Department, and Air Service all had an interest at stake.

In the midst of this legislative give and take, the War Department finally got around to officially issuing Training Regulation No. 440-15, *Fundamental Principles for the Employment of the Air Service,* on 26 January 1926. This was the same training regulation that Patrick had drafted and the Air Service had been using for nearly three years. It, of course, included the admonition that "[t]he organization and training of all air units is based on the fundamental doctrine that their mission is to aid the ground forces to gain decisive success." But, more important, it also accepted Patrick's concept of an independent GHQ air force.[15] The publication of 440-15, at such a critical juncture and hailed by no less an Air Service critic as Maj. Gen. Hugh Drum as "the most advanced thought in the world today on aviation," was enough to give Patrick pause.[16] Even Secretary of War Dwight Davis "insisted that it was very important to pass an air bill through Congress in the spring of 1926 because it was necessary to increase the efficiency of the Air Force and because, in his opinion, the country demanded it."[17]

An interesting sideshow to this legislative brawl occurred in early February when Maj. Hap Arnold, chief of Patrick's headquarters Information Division, and Maj. Bert Dargue, a Patrick favorite (in fact, the one who taught Patrick to fly), took it upon themselves to send off a letter in support of the Patrick/Wainwright measure.[18] Arnold wrote the letter and enlisted Dargue's help to duplicate it with a mimeograph machine in Air Service headquarters. Enclosed with the letter was an outline of the Patrick/Wainwright Bill. These anonymous circulars were forwarded to a variety of individuals, many regular and reserve Air Service officers, some even on Capitol Hill.[19]

Such a brazen attempt to influence legislation met with swift retribution from the General Staff. General Order No. 20 specifically prohibited any attempt to influence legislation except through regular military channels, and the inspector general of the Army was in hot pursuit. So was General Patrick. He appointed Brig. Gen. James Fechet to look into "the alleged secret publication and distribution" of the document.[20] Arnold and Dargue were soon identified as the guilty parties, with Arnold the main conspirator. Patrick was incensed at the whole affair, and, to make matters worse, Billy Mitchell (who had just resigned) used the incident for his own political purposes. Mitchell claimed that the "whole incident was another attempt by the War Department to bludgeon the Chief of the Air Service into silence before Congress" with regard to the passage of legislation.[21] Mitchell made even more fantastic claims, stating that Patrick was "the victim of an espionage system within the department."[22] All of this came about just as Patrick was to testify before the House Military Affairs Committee to recommend the eventual creation of an independent air force. The timing could not have been worse, and Arnold suffered the brunt of Patrick's wrath.

This was an especially unsettling affair for Patrick because he thought highly of Arnold. Patrick himself had posted a letter of commendation to Arnold's personnel file for his work at Rockwell Field and subsequently arranged for his transfer to the Army Industrial College before Arnold assumed the position of chief of information in February 1925.

It has been suggested that Patrick brought charges against Major Arnold in retribution for his quite open support of Billy Mitchell during the recent trial.[23] This was not the case. While Patrick did indeed caution his junior officers to be careful regarding their association with Mitchell while the trial was in progress, this can be attributed to Patrick's genuine concern for their career progression. Patrick witnessed first-hand Mitchell's self-immolation and did not want this to happen to any other of his officers. As noted previously, Patrick had supported Mitchell for reappointment as his deputy while Secretary of War Weeks wanted Mitchell's head in the wake of Mitchell's machinations concerning his series of articles in the *Saturday Evening Post* and his inflammatory testimony to the Lampert Committee. Patrick did not want lackeys or "yes men." What he did want was to gain and retain the respect of the General Staff for his officers and the Air Service as a whole. Arnold was not the only Air Service officer who openly supported Mitchell during the period of the trial. Both Carl "Tooey" Spaatz (later to command the Eighth Air Force in World War II and become Air Force chief of staff, 1947–48) and Eaker, among others, were enthusiastic Mitchell backers.[24] Arnold was not singled out for his support of Mitchell. Arnold's surreptitious middle-of-the-night mailings clearly violated War Department policy, and, according to the incident investigation report, Arnold initially denied everything. This made matters only worse. The timing of the incident, in the midst of a legislative battle and following on the heels of Mitchell's trial, intensified Patrick's ire. Last, so public a violation—on Capitol Hill no less—virtually forced Patrick to take a harsh stand.

Patrick offered Arnold the choice of a court-martial or resignation from the Army.[25] According to Arnold, after consulting with his wife, Bee, and possibly a friend from the judge advocate general's office, he called Patrick's bluff by accepting the court-martial. Now it was Patrick who was in a tight spot. Patrick did not want another public court-martial.[26] He backed down from his initial threat and banished Arnold to Fort Riley, Kansas, where he stayed until Patrick was long retired. It was an exile that Arnold never forgave or forgot.

Patrick was faced with salvaging what he could of the legislation he had labored so long to realize. But with all contending parties desiring legislation and willing to compromise, the committees set to work. It took the House Committee until 29 March to hammer out *An Act to Provide More Effectively for the National Defense by Increasing the Efficiency of the Air Corps of the Army of the United States, and for Other Purposes.*[27] Based primarily on the Morrow Board, it gave little to the Air Service except a new name, and a five-year program for more personnel and equipment, but no plan for appropriations. It was referred to the Senate on 6 May.[28] In the Senate the pro-Morrow Wadsworth Bill formed the basis of what ultimately was combined with the House bill to form the Air Corps Act that was sent to the President and became law on 2 July 1926.[29]

The new law gave the Air Service five reforms.[30] The first and most obvious was the name change to the Air Corps. This was not a casual entitlement, but an acknowledgment of the independent capability of the Air Service that Patrick fought so very hard for. This was a doctrinal victory in a skirmish that Patrick had started when he reorganized his headquarters staff in December 1921, making the Air Service headquarters responsible for doctrinal development.[31]

The second result of the Air Corps Act was the establishment of an additional assistant secretary for war who would be responsible for Army aviation matters. This was more than a matter of prestige, for it created direct representation of aviation matters within the secretary of war's office. And to further increase the Air Corps representation and influence on the General Staff, an Air Corps flying officer would be posted to each General Staff office.[32]

The third major issue of the act also dealt with additional personnel. In this provision, the Air Corps chief would receive two additional brigadier generals. One would head the newly created Materiel Division at McCook Field, and the other would be the commander of the Air Corps Training Center at San Antonio.[33] By way of personnel actions, the act directed that "at least 90 per centum of the officers in each grade below that of brigadier general shall be flying officers . . . and flying units in all cases [will] be commanded by flying officers."[34]

Fourth, the bill made into law the ability to negotiate for new planes without holding a design competition, aircraft procurement practices that up until that time were administered at the discretion of the secretary of war.

The last and final portion of the bill dealt with a five-year expansion program in equipment and personnel. This part of the legislation, more than any other, was critical if the Air Corps was to begin replacing their obsolescent equipment. While the Air Service was currently authorized 1,254 aircraft, the act increased the active inventory to 1,800. While this force was much smaller than Patrick wanted, it was a start. But it was questionable whether the new planes would start to flow at all, due to the way the provision was worded. The president could delay, at his discretion, the necessary aircraft appropriations.[35] As far as personnel increases were concerned, 403 officers and 6,240 enlisted would be added.[36]

Victory or Defeat

The Air Corps Act was a political compromise that effectively ended the major debates concerning Air Service independence for the next eight years.[37] There was still sniping, primarily from Mitchell, mostly in the form of magazine articles. But Mitchell's influence was not what it had once been. His last book, *Skyways,* published in 1930, sold nowhere near the copies that *Winged Defense* did when it was published in August 1925. In the midst of the Depression, few people worried about this aspect of national defense.

One could say that the War Department won this legislative round, as the Air Corps Act was the Morrow Board recommendations made into law. But the War Department did not enter into this arrangement of its own free will. The War Department was nudged in this direction by political considerations, primarily by a president who appointed a blue-ribbon panel to "offer a resolution" concerning the Air Service problem. The Air Corps Act came about because the executive and legislative branches deemed it necessary. Patrick was in a position to take advantage of that political necessity. And he did so, within the system.

Patrick was pragmatic about the results of the legislative battle, knowing that the Air Service received only that which was realistically achievable. In several ways, the Air Corps Act provided much more than one would believe possible given the political, economic, and world situation at the time. Patrick himself said as much, noting that as nations were reducing their armaments, it could hardly be expected that the Air Service could expand and modernize to any greater degree than that which was allowed for in the Air Corps Act.[38]

Patrick made the best of what the Air Corps Act of 1926 had to give. On the surface it actually provided some resolve, if not resolution, to some vexing problems. The creation of the Air Corps was the implementation of Patrick's transitional phase to the realization of an independent air force. The establishment of the Air Corps was the acceptance and acknowledgment of the independent missions the Air Corps could perform. To accomplish those missions, Patrick gained an increase in personnel and aircraft. Many of those personnel would become part of the General Staff and provide critical Air Corps representation at that level on a daily basis. Additionally, those Air Service officers would be pilots. The addition of a second assistant secretary of war considerably raised the visibility level of Air Corps interests. The two brigadier general command positions increased the importance of the materiel and training centers as well.

The five-year aircraft acquisition program was, by far, the greatest aspect of the Air Corps Act. Aircraft modernization not only provided the Air Corps with safer planes of higher performance, but it provided a lifeline to the aircraft manufacturing industry. Patrick had been attempting to implement this goal since 1921. He worked diligently to achieve it with the Lassiter Board but saw it frustratingly ignored in the political skirmishing of the Joint Board. Patrick also realized an additional victory with the modifications in the rules regarding aircraft design competitions. The legislation finally codified what Patrick had implemented in 1924, that the government would not be in competition with private industry. With regard to signing contracts, Patrick could not easily implement a sole-source contract, but the final results did give him some loopholes, a step in the right direction.[39]

Patrick would meet with further frustration in the actual implementation of the Air Corps Act. He was due for mandatory retirement in November 1927 and therefore had about seventeen months to begin implementation. In reality, twelve months of that time were taken up as various studies determined the particulars. While the studies took place, Patrick readily made administrative changes to accommodate the act's requirements. When it came to the actual implementation of the five-year acquisition program, Patrick experienced problems that were still unresolved when he retired, problems that continued five years later for his successors. The problems, though, would have been much worse, if the Air Corps Act had not come into being. But the year 1926 was important not only for the passage of the Air Corps Act. It was a watershed in another way.

The Air Service and aviation technology in general had been held back by lack of funding and an atmosphere of isolationism and pacifism. In addition, the war stocks of leftover aircraft and replacement Liberty engines had chained the Air Service to obsolete technology for eight years. It was only in 1926 that the World War I aviation related stocks were depleted. They had either been used up, declared unsafe or obsolete, or were destroyed. This was very welcome news to the ten aircraft companies that actually survived the drought.

This situation presented the ideal environment in which President Coolidge could implement the provisions of the five-year acquisition program. He did not. This opportunity was ignored even though provision was made in the act for a supplemental budgetary authorization, on the part of the executive branch, to begin the program in the new fiscal year, which began the day before the bill was approved (2 July 1926). Because of this lapse, the five-year program did not begin until the next fiscal year, beginning 1 July 1927. President Coolidge and "his administration took full advantage of the merely permissive, rather than mandatory, character of the legislation."[40] Funding for personnel and aircraft increases was deferred. The one option left open to satisfy the funding and manpower requirements of the new Air Corps were the other agencies of the War Department.[41] Patrick did not want this to happen, and did not envision that it would.

12. Conclusion

> I want to put you at the head of it and have you bring order out of what is now chaos, have you manage it and get results.
>
> General John J. Pershing, Commander, AEF, May 1918
>
> General Patrick repeatedly had to exercise all the tact and diplomacy at his command. His task was never an easy one but his energy and ability enabled him to overcome obstacles which others less able than he might never have been able to circumvent.
>
> F. Trubee Davison, Assistant Secretary of War in Charge of Aviation, May 1928

To understand this era with regard to aviation and Air Service development, one must have a good understanding of the personalities involved, who they were and what they said at specific times, to appreciate the evolutionary thinking of Mason Patrick. Because, quite simply, Patrick changed his mind. Patrick, though, was not the only one who changed his beliefs about the role and place of military aviation. In addition, there was a perceptible paradigm shift in how America as a whole viewed aviation.[1] Some of the credit for that change in the American public psyche goes to Billy Mitchell, but the credit for shifting the intellectual and philosophical opinions of the critically important members of the General Staff, and members of Congress, belongs to Mason Patrick. Patrick not only deftly managed to sway General Staff opinion, but he was also critical in getting the message out to the professional military education institutions and professional civilian and business organizations as well.² Specific turning points in this evolutionary process ultimately culminated in Patrick believing in the future promise and capabilities of air power much as Mitchell did. The fact that Patrick became not only an airpower advocate but also a true believer in an independent air force demonstrates surprising intellectual and philosophical conversion. It was a transformation that began during his days in France and became complete within two years after he again took over as chief of the Air Service in 1921.

When Mason Patrick's six-year battle is compared to the cumulative result of the Air Corps Act, it may seem that his efforts resulted in little. But that is only a superficial appraisal because many who view Patrick's efforts evaluate the results from the perspective of Billy Mitchell's biographers. Immediate independence was not the issue for Patrick; the survival of the Air Service in the crisis days of World War I was. Then it was the survival of the Air Service upon the resignation of General Menoher in 1921, followed by the air arms rehabilitation and eventual semiautonomous status.

Patrick steered the Air Service through a revolutionary period when technology seemed to offer the promise of a new independently decisive weapon but, in reality, delivered much less. Patrick was confronted by a General Staff that won a war on the ground: the Aisne-Marne, St. Mihiel, and Meuse-Argonne campaigns of 1918. Aircraft had assisted with those victories, but aircraft had not won them.

From the start, Patrick was confronted with emotional and doctrinal issues. In 1921, when he once again took over the Air Service, funding was an issue as well. But Patrick had two significant characteristics that helped him face these challenges: a talent for administration and the ability to appreciate new concepts and adapt to them. He had a third significant advantage: he was a well-known part of the military hierarchy and establishment.

Using these talents, Patrick reached out to establish the three legs of his aviation triangle: military aviation, commercial aviation, and the aviation manufacturing base. He initially got his own house in order and put Mitchell on the straight and narrow. Patrick then went on to construct a wide-ranging base of support among War Department, federal, commercial, professional and public interests. He was, arguably, the most active proponent of civil aviation development during his time as Air Service chief. As such, he gained the support of critical allies in his quest to move his legislative agenda through Congress. This included action and legislation to not only maintain the nation's aviation manufacturing base but ultimately to make it flourish.

Patrick's most important accomplishment, though, was on the intellectual level. And this was significant in two ways. Initially, at the end of World War I, Patrick was firmly against an independent air arm. His transformation with regard to the capabilities of air power demonstrated a significant capacity for intellectual observation and evaluation. This ability was the genesis of a renewed vision for the Air Service. While General Menoher stubbornly attempted to constrain the promise of this new combat arm, General Patrick became convinced of its potential and cultivated its capabilities.

Patrick did this by persuasive and persistent advocacy. He convinced the General Staff and War Department of the current and ultimate value of air power, significantly, air power based on War Department–sanctioned doctrine. Patrick's role concerning the formal codification of that airpower doctrine has been long overlooked. He was responsible for the first statement of airpower doctrine accepted by the War Department. Patrick's belief in and promotion of "Fundamental Conceptions" was the bedrock of the intellectual acceptance by the General Staff concerning the independent capabilities of the air combat arm. Once this hurdle was overcome, Air Service expansion and ultimate independence were much easier to justify. That justification came with the signing of the Air Corps Act of 1926.

Personally, Patrick was not known as an ebullient or outgoing personality. But at the same time he was not the dour demon—the enemy of Billy Mitchell—that some portray him to be. One must look broadly and deeply to find the persona of Patrick.

He was an honest man and an honorable one. He knew his limitations but was not afraid to test their boundaries (witness his successful pilot training). Limitations in this case apply to what was realistically achievable given his environment. When the environment changed, politically, economically, and individually, as it did from 1921 to 1927, Patrick was quick to adapt to those changes and take advantage of them. He did not force change as much as he gently molded it to fit opportune circumstance over time. When complications arose, he did not retreat, but outflanked the obstacle, such as his goals regarding manufacturing incentives, commercial air development, and the Air Corps Act. He stood firm when required, handling Billy Mitchell being a prime example of his strength.

Reflecting on Patrick's time in uniform, which spanned forty-two years of service, one is struck by the major accomplishments in the twilight of a long and eventful career. At the age of fifty-seven Patrick took on the challenge of commanding the Air Service, albeit reluctantly when offered, but enthusiastically when confirmed. He displayed amazing stamina, supreme patience, and intellectual depth contending with a wide variety of contentious issues. Challenged by his second in command the first day on the job, Patrick finessed the issue with ease. But he was not vindictive, regardless of Hap Arnold's personal invective in later years. Arnold was the only officer who served under Patrick to hold him in low esteem, and this opinion only came about because of Arnold's provocative action. Even Bert Dargue, who was punished along with Arnold for the 1925 circular mail prank, never criticized Patrick. Arnold, though, in later years, doubtless modified his harsh views based on his own experiences.

Then there was Billy Mitchell. Mitchell never criticized Patrick directly. He criticized Patrick's programs, the Lassiter Report, the Patrick Bill, and the Air Corps Act for not going far enough, for not demanding air force independence. Patrick used Mitchell's dynamic personality, intelligence, and vision to complement his own approach to Air Service problems. But Patrick was the more rational and far-sighted. No one can say that Patrick was not a visionary. There were times when Patrick outdistanced his brash assistant in forecasting the future potential of air power. Drawing upon Liddell Hart's vision of armed conflict in *Paris: Or the Future of War,* Patrick foresaw that air power could be a decisive weapon of war.[3] "Even Mitchell hesitated to say that air power could be decisive without ground and naval action."[4]

Patrick was a man of vision. Mitchell was a man of vision, but it was a vision with self-inflicted limitations. Mitchell's vision was certainly unfettered with regard to the promise of air power, but it was circumscribed by a tunnel-vision approach dictating immediate and radical change. Mitchell completely dismissed the chance for a realistic appraisal of what was possible given the current environment. He would not compromise and, in doing so, sacrificed his career. Mitchell's very public obstinacy promoted an even more inflexible response from the traditionalist General Staff and War Department. One good thing did come out of Mitchell's rather uncooperative attitude in 1921, and that was Menoher's resignation, which led to Patrick's new assignment, which Pershing had doubtless engineered far in advance.

Mitchell's approach to the control of the civil aviation transportation sector was flawed. While both Mitchell and Patrick saw the development of civil aviation as integral to national security, Mitchell proposed putting the administration of civil aviation under the military. It was an idea that did not warrant serious consideration and even engendered incredulous response.

Patrick's approach to improving both the civil and military aviation communities was a cogent all-encompassing solution to a multifaceted problem. Utilizing the administrative talent that Pershing knew he possessed in abundance, Patrick quickly envisioned what he must do to rectify the many problems before him. The ultimate solution to those problems was diametrically opposed to the beliefs of many of Patrick's personal and professional acquaintances. He did not let this deter him from what he knew to be the correct course of action, and the most appropriate, given the circumstances of the time.

In the twilight of his tenure as chief of the Air Corps, Patrick received a lengthy letter from an individual who, in taking a direct slap at Billy Mitchell's philosophy, emphasized the severe limitations of military aviation.[5] Patrick, in reply, chastised the writer for stressing "so energetically this negative point of view." At the same time, Patrick sympathized with the writer: "Of course, I know that you are merely combating what you consider to be exaggerated and pernicious propaganda, and with this I have no fault, much harm has been done by extravagant statements of the power and capabilities of [air] warfare."[6] At the same time, Patrick noted: "There is nothing to be gained by minimizing [air warfare's] great strategic possibilities . . . in the importance of superiority on the field of battle we should not neglect the opportunities for decisive results through indirect [strategic air warfare] methods."[7]

The fact that Patrick did not always believe as such is a testament to his intellectual flexibility. The fact that Patrick persuaded key decisionmakers to see the merits of his arguments and act upon them is even more significant.

The journey from the AEF Air Service days of 1918 to the semi-autonomous Air Corps of 1927 witnessed a radical change in the concepts of air warfare in the United States. Patrick was directly responsible for a large part of that transformation. It was a transformation that put the U.S. Air Service on the road to independence.

It is appropriate to repeat what General Pershing said to Patrick the day the AEF commander offered his West Point classmate the position as chief of the AEF Air Service: "I am convinced that you are the best one I can find for this job." He was right.

The Air Corps Act was the capstone of Patrick's career, and for the remaining months of his tenure he was concerned primarily with the implementation of the act's five-year expansion program. The Coolidge administration, though, except for implementing the cosmetic aspects of the legislation, was not an enthusiastic partner in this regard. Air Service budgets and aircraft acquisition improved, but due primarily to Patrick's energetic prodding.

Patrick kept busy in his support of all things aviation related. The Air Commerce Act, which Patrick had endorsed on numerous occasions, was passed in May 1926, and he continued to speak to a wide variety of military, public and, private audiences about America's need for preeminence in aviation. His retirement on 13 December 1927, after forty-one years of active duty, was followed by the publication of his 1928 book, *The United States in the Air,* which covered his involvement in American aviation and made a strong case for a more dynamic and expansive aviation policy for the nation. Remaining in Washington, D.C., at the house where he had lived since 1922, he kept busy on the lecture circuit, especially at military and civilian schools and to civic groups. In June 1929 Patrick was appointed public utilities commissioner of the District of Columbia by President Hoover and served in this capacity until 1933. Patrick's wife, Grace, died several years later. In early December 1941, afflicted with heart disease and cancer, Patrick was admitted to Walter Reed Hospital in Washington and died on 29 January 1942 at the age of 78. His adopted son, Bream, a captain in the Air Corps, was at his side. Patrick was buried in Arlington National Cemetery. On 26 August 1950 the U.S. Air Force renamed their air weapons proving ground base at Cocoa, Florida, Patrick Air Force Base.[8]

Within the numerous condolence letters sent to Bream Patrick following his father's death was a brief typewritten letter from the War Department Office of the Deputy Chief of Staff for Air, the individual who was in charge of the Army Air Forces, the organization that replaced the Army Air Corps:[9]

The passing of your father was a severe loss to the Army Air Forces and to the nation he served so well. I wish to add my respects to the memory of our former leader and a great American.

With prophetic wisdom and rare tenacity of purpose your father dedicated his life to the development of aviation and the defense of our country. We of the Air Forces, engaged in a wartime struggle for aerial supremacy, fully realize that we are building on the foundation which he so laboriously constructed. The expanding Air Forces stand as a monument to General Patrick's vision and genius.

Expressing to you my deep and sincere sympathy, I am,

Sincerely yours,

(signed)

H. H. ARNOLD

Lieutenant General, U. S. Army

Given the rocky relationship between these two great airmen, one can ask whether Arnold's words were sincere, but considering how far the Air Service and Hap Arnold had come since 1925, it is a good bet that Arnold meant what he said. For Arnold was, indeed, building on the foundation of Patrick's vision and genius.

Notes

Abbreviations

ACS	Assistant chief of staff
AFHRA	Air Force Historical Research Agency, Maxwell Air Force Base, Montgomery, Ala.
AFHSO	Air Force History Support Office, Washington, D.C.
CoAS	Chief of the Air Service
LC	Library of Congress, Washington, D.C.
MHI	U.S. Army Military History Institute, Carlisle Barracks, Carlisle, Pa.
MMP	Mason M. Patrick
NA	National Archives, Washington, D.C.
OAFH	Office of Air Force History, Washington, D.C.
PRO	Public Record Office, London
RFC	Royal Flying Corps
RG	Record Group
USAFA/SC	USAF Academy Library, Special Collection Branch
USMA	U.S. Military Academy

Introduction

1. Charles J. Gross, "George Owen Squier and the Origins of American Military Aviation," *Journal of Military History* (July 1990): 288.
2. Maurer Maurer, *Aviation in the U.S. Army, 1919–1939* (Washington, D.C.: OAFH, 1987), xxi–xxii.

3. Isaac Don Levine, *Mitchell, Pioneer of Air Power* (New York: Duell, Sloan and Pearce, 1958), 327.
4. In addition to Hurley's book, also see ibid., and Michael L. Grumelli, "Trial of Faith: The Dissent and Court-martial of Billy Mitchell," Ph.D. diss., Rutgers University, New Brunswick, N.J., 1991.

Chapter 1. From the Wright Brothers to World War I

1. Alfred Goldberg, ed., *A History of the United States Air Force, 1907–1957* (Princeton, N.J.: D. Van Nostrand, 1957), 2. In 1914 Glenn Curtiss, noted aviator and entrepreneur, made major modifications—in fact, rebuilt—Langley's "Aerodrome A" into a flyable airplane based on the Wright brothers' technology. See Charles H. Gibbs-Smith, *A History of Flying* (London: Batsford, 1953), 209–12.
2. Raymond R. Flugel, "United States Air Power Doctrine: A Study of the Influence of William Mitchell and Giulio Douhet at the Air Corps Tactical School, 1921–1935," Ph.D. diss., University of Oklahoma, Tulsa, 1965, 10. See also Lee Kennett, *The First Air War, 1914–1918* (New York: Free Press, 1991), 20–22.
3. See Michael Adas, *Machines as the Measure of Men: Science, Technology, and Ideologies of Western Dominance* (Ithaca, N.Y.: Cornell University Press, 1989); and Wolfgang Schivelbusch, *The Railroad Journey: The Industrialization of Time and Space in the Nineteenth Century* (Berkeley: University of California Press, 1986).
4. "List of Appropriations for Army Aviation, Army, Navy Aviation, Navy, Combined Services, 1890–1939," n.d., File 131.41-2, Budget and Fiscal Division, Army Air Forces, AFHRA; also Goldberg, ed., *History of the United States Air Force*, 2.
5. Juliette Hennessy, *The United States Army Air Arm, April 1861 to April 1917* (Washington, D.C.: OAFH, 1985), 59.
6. Chief Signal Officer, memo to Chief of Staff, 28 March 1913, in Central Files 421A-Insignia, and CSO, memo to Chief of Staff, in Central Files No. 211-Aviators, both in RG 407, Adjutant General Office, National Archives, Washington, D.C. (hereafter NA); also in Hennessy, *United States Army Air Arm*.
7. David MacIsaac, "The Air Force," in *Encyclopedia of the American Military*, ed. John E. Jessup and Louise B. Katz (New York: Scribner's, 1994), 422. See Hennessy, *United States Army Air Arm*, 136–45, for a detailed discussion of the San Diego Signal Corps Aviation School.
8. Hennessy, *United States Army Air Arm*, 167–76.
9. War Department bookkeeping for the total number of aircraft purchased from 1909 to 1917 is anything but specific. The totals vary from 165 to over 400, depending on the source. Hennessy discusses same in *United States Army Air Arm*, 196–97. Even H. H. "Hap" Arnold, in *Global Mission* (New York: Harper, 1949), mistakenly notes that there were fifty-five airplanes in the Aviation Section inventory when the United States entered the war.

10. Arnold, in *Global Mission,* 50, notes that there were fifty-five officers (to include twenty-six qualified pilots) assigned to the Aviation Section when the United States entered the war. He was technically correct but had the "big picture" wrong: there were fifty-five Regular Army officers detailed to or assigned to the Aviation Section, but there were, *in toto,* 131 qualified military aviator officers on the books, spread throughout the Army. See also Hennessy, *United States Army Air Arm,* 196.
11. Mason M. Patrick, *The United States in the Air* (Garden City, N.Y.: Doubleday, Doran, 1928), 49. Given the lack of personal papers associated with Patrick, his one and only book provides excellent background and chronology and assisted greatly in this study.
12. Alfred F. Hurley and William C. Heimdahl, "The Roots of U.S. Military Aviation," in *Winged Shield, Winged Sword: A History of the United States Air Force,* ed. Bernard C. Nalty (Washington, D.C.: Air Force History and Museums Program, 1997), 1:10.
13. Ibid., 27–28.
14. See Hennessy, *United States Army Air Arm,* 93, for a description of the four aircraft-control systems of the time.
15. Hurley and Heimdahl, "Roots of U.S. Military Aviation," 11.
16. James J. Hudson, *Hostile Skies* (Syracuse, N.Y.: Syracuse University Press, 1968), 3.
17. Richard P. Hallion, *Strike from the Sky: The History of Battlefield Air Attack, 1911–1945* (Washington, D.C.: Smithsonian Institution Press, 1989), 10.
18. *United States Army in the World War, 1917–1919, Reports of the Commander-in-Chief, Staff Sections and Services* (Washington, D.C.: Center of Military History, U.S. Army, 1991), 15:236 (hereafter *Reports of the Commander-in-Chief*); Burke Davis, *The Billy Mitchell Affair* (New York: Random House, 1967), 30.
19. See Williamson Murray, "Strategic Bombing: The British, American, and German Experiences," and Richard Muller, "Close Air Support: The German, British, and American Experiences, 1918–1941," in *Military Innovation in the Interwar Period,* ed. Williamson Murray and Allan R. Millett (Cambridge: Cambridge University Press, 1996); John H. Morrow, *German Air Power in World War I* (Lincoln: University of Nebraska Press, 1982); as well as Hudson, *Hostile Skies.*
20. Richard Hallion, *Rise of the Fighter Aircraft* (Baltimore, Md.: Nautical and Aviation Publishing of America, 1988), chap. 1 passim. Also unpublished lecture by Dr. Richard Hallion, "World War I Aviation," on file at Office of the Air Force Historian, Bolling Air Force Base, Washington, D.C.
21. Davis, *Billy Mitchell Affair,* 28–29.
22. Ibid., 37.
23. William Mitchell, *Memoirs of World War I: From Start to Finish of Our Greatest War* (reprint: Westport, Conn.: Greenwood Press, 1975), 177–78. Roger Burlingame, in his book, *General Billy Mitchell* (New York: Greenwood Press, 1972), 74, notes that at this time "Mitchell had [nothing] against Foulois," but this is belied by what

Mitchell both wrote and said in contemporary and future accounts, and the fact that Mitchell and Foulois had a major run-in during the Mexican Punitive Expedition.

24. Lt. B. D. Foulois, "Provisional Aeroplane Regulations," Foulois Papers, MS 17, Topical Files, Box 17, Series 5, USAFA/SC.
25. MS 17, Manuscripts, Box 6, Series 2, Folder 6, USAFA/SC, 178. As with a number of other items supposedly written by Mitchell, this quote was appropriated from another officer. See Grumelli, "Trial of Faith."
26. William Mitchell, letter to Ruth Mitchell, 23 Nov. 1917, MS 17, Manuscripts, Box 6, Series 2, Folder 6, USAFA/SC; Davis, *Billy Mitchell Affair,* 35.
27. Pershing's action to separate aviation from the Signal Corps led many in the Air Service to think that Pershing would support their future drive for independence, but such was not the case. See Robert Frank Futrell, *Ideas, Concepts, and Doctrine: Basic Thinking in the United States Air Force, 1907–1960* (Maxwell AFB, Ala.: Air University Press, 1989), 35.
28. John J. Pershing, *My Experiences in the World War* (New York: Frederick A. Stokes, 1931), 161; Futrell, *Ideas, Concepts, and Doctrine,* 21.
29. Pershing to War Department, cable no. 659-S, 28 Feb. 1918, Harbord Papers, as cited in manuscript, "His Prophetic Vision Soared Away Into Space" (author unknown), AFHSO.
30. Wesley Frank Craven and James Lea Cate, eds., *Plans and Early Operations, January 1939–August 1942,* vol. 1 of *The Army Air Forces in World War II* (Chicago: University of Chicago Press, 1948), 9; "His Prophetic Vision," 46–47. Also see Daniel R. Mortensen, "The Air Service in the Great War," in *Winged Shield, Winged Sword,* ed. Nalty, 1:50–51.
31. Craven and Cate, eds., *Plans and Early Operations,* 9. Also see Pershing, *My Experiences in the World War,* 333–34.
32. Mortensen, "Air Service in the Great War," 51. Without a civilian second assistant secretary to represent Air Service interests to the secretary of war, after the war the new combat arm was at a decided disadvantage. This disability continued to haunt the Air Service, and especially Mason Patrick, well into the 1920s. See Patrick, *United States in the Air,* 9.
33. See Paul F. Braim, *The Test of Battle: The American Expeditionary Force in the Meuse-Argonne Campaign* (Newark, N.J.: White Mane Publishers, 1983), 158.
34. Patrick, *United States in the Air,* 7.

Chapter 2. Army Engineer from the Hills of West Virginia

1. MMP, letter to John J. Mapel, 1 Aug. 1924, in Records of the Office of the Chief of the Air Corps, Maj. Gen. M. M. Patrick, Correspondence, 1922–1927, Entry 228, Box 5, RG 18, NA (hereafter Patrick correspondence).

2. Bruce A. Bingle, "Building the Foundation: Major General Mason Patrick and the Army Air Arm, 1921–1927," M.A. thesis, Ohio State University, Columbus, 1981, 6–7. Details of Patrick's early life taken from an address delivered by Patrick at the annual Convention of the Daughters of the Confederacy, 18 Nov. 1926, Richmond, Va., Entry 229, Box 2, RG 18, NA.
3. Patrick loved a fine cigar. A 1925 letter he fired off to a tobacconist attests to this (and to his frank manner of speaking as well): "I received [your cigars] and have smoked one. This is enough for me. These are the worst cigars you have ever sent me. Replace with my check, or something I can smoke." Patrick, letter to Thompson and Co., 5 Aug. 1926, Records of Chief of Air Corps, Entry 228, Box 7, RG 18, NA.
4. *Beckley* (W.V.) *Post-Herald,* 1 Sept. 1978, 12; Roger J. Spiller, ed., *Dictionary of American Military Biography* (Westport, Conn.: Greenwood Press, 1984), 2:826–29; *Lewisburg* (W.V.) *Gazette,* 2 July 1978, 2; Mason M. Patrick File Card, Class of 1886, Cadet Personnel Records, USMA Archives, West Point, N.Y.
5. Register of Delinquencies, 22:259, 421, 530, USMA Archives.
6. Patrick, *United States in the Air,* 5. Pershing was twenty-one years and nine months old upon his admission to West Point.
7. USMA, *Post Order Book,* no. 10 (7 July 1880–19 Nov. 1884): 315–16; *Post Order Book,* 2:75; Mason Mathews Patrick, File Card/1886 Class Card, Cadet Service Record, General Merit Roll of 1886, USMA Archives; and "Official Register of the Officers and Cadets of the U.S. Military Academy," West Point, N.Y., June 1886.
8. G. W. Cullum, *Biographical Register of the Officers and Graduates of the U.S. Military Academy at West Point, New York Since Its Establishment in 1802, to 1890,* 3d ed. (New York: R. R. Donnelley and Sons, 1891), 3:392; Myron J. Smith Jr., "Mountaineer Boss of the Eagles," *Lewisburg Gazette,* 2 July 1978, 3; and Bingle, "Building the Foundation," 8.
9. Cullum, *Biographical Register of the Officers and Graduates,* Supplement, 5:379 and 3:392; Patrick, letter to Capt. Theodore S. Cox, 31 March 1926, Entry 228, Box 1, RG 18, NA; Maj. Gen. Fred Sladen, West Point Superintendent, letter to Patrick, 6 Feb. 1925, Records of the Office of the Chief of the Air Corps, Correspondence 1922–27 (R-S), Entry 228, Box 7, RG 18, NA.
10. MMP, letter to Mrs. Louis Bennett, New York, N.Y., 29 Oct. 1919, Louis Bennett Family Papers, R65E-2, Box 2, File 2, Special Collections, West Virginia University, Morgantown (hereafter Bennett Collection).
11. As Patrick's service record (201 File) was destroyed in the 1973 National Personnel Records Center fire, his assignment history was gleaned from information available at the USMA archives and library. In particular: "School History of Candidates," no. 1, 1880–99, USMA Archives; and the individual USMA Cadet Biographical Files, USMA Library.
12. "Statement of Military Service," USMA Cadet Biographical Files: Mason M. Patrick,

1; Scrapbook of Papers, in Miscellaneous Papers of Maj. Gen. Mason M. Patrick, U.S. Army Aviation Museum, Fort Rucker Army Base, Dothan, Ala. (hereafter Fort Rucker Scrapbook).
13. Maj. Gen. Oscar Westover, "The Distinguished Personal and Military Record of Major General Mason M. Patrick," File 168.7089-50, pp. 3–4, AFHRA.
14. Pershing explained to Patrick that Blatchford "had not grasped the problems confronting the Commanding General, that there were heads which he must cut off and that he was going to put me in that place temporarily." See entry for 31 Oct. 1917 in Patrick diary, Patrick Papers, SMS 198, USAFA/SC (hereafter Patrick diary). (Pershing did this after discussing S.O.S. problems with Patrick after an S.O.S. conference; Pershing asked Patrick to remain and talk after the formal meeting was over. The change of command orders were cut that night.)
15. Donald Smythe, *Pershing: General of the Armies* (Bloomington: Indiana University Press, 1986), 82; Patrick diary, 20–23 Feb. 1918.
16. Patrick, *United States in the Air,* 4–5.
17. MMP, letter to Grace Patrick, 11 May 1918, Fort Rucker Scrapbook.
18. Patrick diaries, 6 Aug. 1917–28 May 1918, 8 Jan.–14 July 1919. Entry for 6–16 May 1918, 82.
19. Patrick, *United States in the Air,* 6.
20. Patrick diary, 6–16 May 1918, 82; and Patrick, *United States in the Air,* 7.
21. Goldberg, ed., *History of the United States Air Force,* 15–16, 22.
22. Pershing, *My Experiences in the World War,* 333; Smythe, *Pershing,* 143.
23. Mitchell, *Memoirs of World War I,* 165–66. Also see Maurer Maurer, ed., *The U.S. Air Service in World War I* ("Gorrell's History"), 4 vols. (Washington, D.C.: OAFH, 1978–79), 1:65.
24. James Hudson, in *Hostile Skies,* is sympathetic, and rightly so, to Foulois's plight in addressing the use of ground officers. As Hudson notes: "there simply were not enough capable aviators available for the gigantic task ahead," and "there is no evidence that [Foulois] ever planned to put a nonflying officer over combat units" (55).
25. For an exceptional account of critical periods in the career of Benny Foulois, see John F. Shiner, *Foulois and the U.S. Army Air Corps, 1931–1935* (Washington, D.C.: OAFH, 1983). The book is far more broad and informative than the title indicates. In addition, Foulois's 2,000-page handwritten manuscript for his book, *From the Wright Brothers to the Astronauts: The Memoirs of Benjamin Foulois* (coauthored with Carroll V. Glines), is on file at the USAF Academy with other important Foulois papers.
26. Futrell, *Ideas, Concepts, and Doctrine;* and Craven and Cate, eds., *Plans and Early Operations,* 10–11, as cited in Shiner, *Foulois and the U.S. Army Air Corps,* 9.
27. Patrick, *United States in the Air,* 7.
28. Craven and Cate, eds., *Plans and Early Operations,* 5, 9.

29. Shiner, "Early and Interwar Years, 1907–1939," in *Winged Shield, Winged Sword*, ed. Nalty, 1:60; H. A. Toulmin Jr., *Air Service, A.E.F., 1918* (New York, 1927), chap. 6.
30. See Army Air Forces, "Organization File," File 321.9A, Box 480, RG 18, NA.
31. Patrick, in his book, *The United States in the Air*, mistakenly cites the total aviation acquisition program as 358 squadrons (p. 17) vice 386. See "Schedule of Personnel and Materials Required to Meet Air Service Program of 202 Squadrons," Air Service Records, Microfiche File 167.404, Roll A1538, Frames 1160–1240, USAF History Support Office, Washington, D.C. For a concise overview of this period, see Goldberg, ed., *History of the United States Air Force*, 15–27; for a detailed official history, consult Lucien Thayer, *America's First Eagles: The Official History of the U.S. Air Service, A.E.F., 1917–1918* (San Jose, Calif.: Bender Publishing, 1983), and *Reports of the Commander-in-Chief*, 245; and for an imminently entertaining read, see Hudson's *Hostile Skies*.
32. Pershing, letter to Patrick, 23 May 1918, and certificate, "Promotion to Major General," both in Fort Rucker Scrapbook; Patrick, letter to Pershing, 27 May 1918, Pershing Papers, Manuscript Collection, Library of Congress, Washington, D.C.; Patrick, memo to Chief of Staff, subj.: "Report on Air Service, A.E.F.," 31 May 1918, quoted in Toulmin, *Air Service, A.E.F.*, 84–91.
33. Patrick diary, 23 May 1918, 82, 84, 86; and William Mitchell, *Memoirs of World War I* (reprint; New York: Random House, 1960), 205. This book was originally printed in serial form in *Liberty* magazine, March–May 1928. The manuscript, which Mitchell completed in 1928, includes a very positive characterization of Patrick, both as an engineering officer and as commander of the Air Service, AEF.
34. Foulois, letter to Pershing, "Relief of Colonel William Mitchell, Air Service," dated 4 June 1918, in Foulois Papers, MS 17, Box 7, Manuscripts, USAFA/SC.
35. For an account of this incident, see Hudson, *Hostile Skies*, 57–58.
36. Major General McAndrews, Pershing's chief of staff, letter to Brigadier General Foulois, 8 June 1918, MS 17, Box 7, USAFA/SC.
37. See "Co-ordination of Air Service Matters," Craig, memorandum to Mitchell, 1 July 1918, in Foulois Papers, MS 17, Box 7, USAFA/SC.
38. Edward Coffman, *The War to End All Wars* (New York: Oxford University Press, 1968), 205.
39. See Patrick diary, entry for 24 March 1919. Looking toward the future, it appears Mitchell's outburst against Foulois and his penchant for exceeding his command authority were portents of his postwar dealings. Obdurate and even "childish" (as noted by Foulois) Mitchell's recalcitrant way of dealing with those in authority seems to have been formed at this juncture.
40. Shiner, *Foulois and the U.S. Army Air Corps*, 9. It is not clear why Foulois requested this action (Foulois never directly addressed the issue), but it is only conjecture that

Patrick was behind the initiative, as this was his original plan that he briefed to Pershing shortly after his appointment as Air Service chief. See Patrick, *United States in the Air,* 8.

41. Patrick, "Final Report of Chief of Air Service, A.E.F." (hereafter Patrick, "Final Report"), in *Reports of the Commander-in-Chief,* 249–50.
42. Patrick diary, entry for 10 Sept. 1918.
43. Patrick, *United States in the Air,* 33–34.
44. Roger G. Miller, "Keep 'Em Flying: A History of Air Force Logistics from the Mexican Border to the Persian Gulf," 1997, 155, on file in AFHSO.
45. Patrick diaries, passim. Patrick's World War I–era diaries begin the day after he ships out for France, with relatively regular entries until Pershing appoints him as Air Service chief. At that point, entries recount weekly activities and invariably mention the hectic pace of his schedule. Business-oriented and never maudlin, he was simplistically eloquent at times. The entry for Easter Sunday, 31 March 1918: "A lovely day—and the battle rages."
46. "Comparative Record of Fatalities," datecd 10 January 1920, Air Service Records, Microfiche File 167.4-13, Roll A1530, Frame 18, USAF History Support Office, Washington, D.C.
47. Patrick, "Final Report," 234, 262–63.
48. Ibid., 234–35.
49. Holley, as cited in Flugel, "United States Air Power Doctrine," 48.
50. I. B. Holley, *Ideas and Weapons* (reprint; Washington, D.C.: Air Force History and Museums Program, 1983), 159, 177.

Chapter 3. Patrick Wins His War within a War

1. Secretary of War, annual reports (1916–19); and Assistant Chief of Staff, Intelligence, Historical Division (hereafter ACS), *Organization of Military Aeronautics, 1907–1935,* Army Air Forces Historical Study no. 25 (Washington, D.C.: Army Air Forces Historical Office, 1944), 28.
2. Goldberg, ed., *History of the United States Air Force,* 25–28.
3. Shiner, "Early and Interwar Years," 70. Ultimately 740 planes were assigned to American air squadrons, which accounted for just over 10 percent of total Allied aircraft. See Hudson, *Hostile Skies,* 300; and Futrell, *Ideas, Concepts, and Doctrine,* 27.
4. A total of 39 planes were manufactured in the United States in 1912; by 1916, the total was up to 411, being manufactured by several dozen relatively small firms. See Roger Bilstein, *The American Aerospace Industry, from Workshop to Global Enterprise* (New York: Twayne Publishers, 1996), 19.
5. This temporary expedient being the result of the May 1918 Overman Act. See Craven and Cate, eds., *Plans and Early Operations,* chaps. 1–2, passim.

6. Patrick, *United States in the Air,* 16–17.
7. As one of the first to proceed to France, Patrick was initially the chief engineer of the Lines of Communication, as well as the director of all AEF construction and forestry in France.
8. Patrick, *United States in the Air,* 22–24. Both Patrick and Trenchard had problems with London. During one of Patrick's calls upon Trenchard, the RFC commander complained bitterly about the British bombing squadrons that constituted the Independent Air Force. This force, controlled directly by the Air Ministry, was, according to Trenchard, "forced upon him."
9. Patrick diary, 20–22.
10. Patrick, *United States in the Air,* 49–50.
11. Patrick, "Final Report," 225. Also note that there were approximately 600 aircraft in eight U.S. Marine Corps squadrons and the twenty-four U.S. Naval Air Stations in Europe.
12. See USAF, *U.S. Air Service Victory Credits, World War I,* USAF Historical Study no. 133 (Maxwell Air Force Base, Ala.: Historical Research Division, Air University, 1969). Patrick's numbers in *The United States in the Air* ended up being almost identical to the 1960 U.S. Air Force study (he notes 781 enemy airplanes destroyed and 289 U.S. losses).
13. Toulmin, *Air Service, A.E.F.,* 327.
14. Patrick, *United States in the Air,* 24, 29–30. Also see various speeches given by MMP during his tenure as Air Service/Air Corps chief, especially to Leavenworth classes and to professional engineering and industrial association gatherings in "Speeches and Articles of Major General M. M. Patrick, 1926–27," in the records of the Army Air Forces, Entry 229, RG 18, NA.
15. Capt. Laurence S. Kuter, "American Air Power School, Theories versus World War Facts," lecture at the Air Corps Tactical School (ACTS), World War History Course, 1937–38, File 248.20(A), AFHRA. (Kuter was chief of the ACTS Bombardment Section, 1936–37.) Also see Flugel, "United States Air Power Doctrine," 47.
16. Billy Mitchell would claim (much after the war) that he was the major proponent behind the idea of strategic bombing. He was not. (Then) Maj. Edgar Gorrell was the driving force behind the concept as it concerned the Air Service, AEF. Mitchell's ideas on the employment of American World War I air assets were essentially tactical (although, granted, under centralized command), with Mitchell even declaring: "Observation Aviation forms the base of an Air Service." See Col. William Mitchell, "Theoretical Diagram: Use of Aviation," 3 March 1918, Air 1/498/15/319/4, PRO.
17. Patrick was quick to implement Pershing's desires, which, in no uncertain terms, supported the soldier on the ground; but Patrick did not unequivocally reject the concept of strategic bombing. He simply, and correctly, grasped the capabilities of the Air Service as they then were and would be within the coming months. See Maj. Gen.

Patrick to Major Fowler, Paris, 24 July 1918, AIR 1/959/204/5/1036 44064, PRO; and Pershing, *My Experiences in the World War,* 2:337.
18. HQ RFC memorandum, 13 Jan. 1918, 4, AIR 1/925/204/5/812, PRO.
19. Pershing, letter to Trenchard, 6 Feb. 1918, in ibid.
20. HQ RFC memorandum, 13/1/1918, 4, and Patrick, letter to Brigadier General Harts, 1 July 1918, both in ibid.
21. Pershing, letter to Trenchard, 6 Feb. 1918, ibid.
22. Patrick, *United States in the Air,* 16–17.
23. A.C. of S., G-2, G.S., G.H.Q., A.E.F. (Brig. Gen. Nolan), letter to British Mission, G.H.Q., A.E.F., Subject: Information on German Industries, 18 Oct. 1918, and Trenchard to Air Ministry, 28 Oct. 1918, both in AIR 1/1976/204/273/39, PRO.
24. CoAS to President of the Air Council, London, 19 Jan. 1918, "Standard Organization of the Allies Aerial Units," AIR 1/925/204/5/812, PRO.
25. "Joint Note to the Supreme War Council by Its Military Representatives," Joint Note no. 35, *Bombing Air Force,* 3 Aug. 1918, Copy to War Cabinet, Air 1/261, PRO. The Pershing/Bliss relationship is interesting with regard to the strategic bombing issue. As much as Bliss, when he was chief of staff, conceived of himself as the "Assistant Chief of Staff to the Chief of Staff of the A.E.F.," he saw his role as serving Pershing to the best of his ability. If Pershing had not endorsed strategic bombing, Bliss would not have endorsed it either.
26. See James J. Cooke, *The U.S. Air Service in the Great War, 1917–1919* (Westport, Conn.: Praeger, 1998), 225–26.
27. Brig. Gen. William Mitchell, "Tactical Application of Military Aeronautics," 5 Jan. 1919, File 167.4-1, AFHRA; Mitchell, *Winged Defense* (New York: Putnam's and Sons, 1925).
28. See Col. W. Mitchell, CAS, 1st AC, AEF, "U.S. Theory of General Disposition of Aeronautical Elements During an Active Period; and Operational Diagram of Use of Aviation," AIR 1/498/15/319/4, PRO.
29. Philip S. Meilinger, e-mail to author, May 15, 2001; see also Andrew Boyle, *Trenchard* (New York: Norton, 1962).
30. Raymond H. Fredette, *The Sky on Fire: The First Battle of Britain, 1917–1918* (Washington, D.C.: Smithsonian Institution Press, 1991), 218, 221.
31. Boyle, *Trenchard,* 288; Eric Ash, *Sir Frederick Ash and the Air Revolution, 1912–1918* (Portland, Ore.: Frank Cass, 1999). See chaps. 5 and 6 for an excellent discussion of the fundamental role that Sykes played in the establishment of the strategic Independent Air Force. It is interesting to note the parallels between the mission of both Sykes and Patrick in that both had to maintain control of their respective air services, rife with headstrong personalities and political issues.
32. Trenchard to Maj. Gen. F. B. Maurice, DMO War Office, 31 March 1917, AIR 1/2266, PRO; and "A Short Review of the Situation in the Air on the Western Front

and a Consideration of the Part to be Played by the American Aviation," 3, para. 4, AIR 725/97/6, PRO. Also see Patrick, *United States in the Air,* 22. The Trenchard and Mitchell relationship on this particular point is crucial to understanding Mitchell's actions in the coming years. A myth arose that "Boom" Trenchard went to Parliament over the heads of the Army and the Navy. He did not. But Mitchell used this historical inaccuracy to rationalize and justify his own actions in the years prior to his court-martial in 1925. See Fredette, *Sky on Fire,* 226–27.
33. Patrick, *United States in the Air,* 23–24.
34. Patrick to Major Fowler, HQ Air Service, American Troops with British, "Aviation with Divisions," 9 July 1918, and Salmond to Patrick (via Major Fowler), 12 July 1918, both on p. 1, AIR 1/925/204/5/812, PRO.
35. Patrick, *United States in the Air,* 23.
36. *Reports of the Commander-in-Chief,* 250.
37. Hudson, *Hostile Skies,* 60.
38. James J. Cooke, *Pershing and His Generals, Command and Staff in the A.E.F.* (Westport, Conn.: Praeger, 1997), 31.
39. Pershing to Brig. Gen. George O. Squier, Chief Signal Officer in the War Department, 29 July 1917, RG 18, NA, as cited in Cooke, *Pershing and His Generals,* 44 n. 1
40. Alfred F. Hurley, *Billy Mitchell: Crusader for Air Power* (Bloomington: Indiana University Press, 1975), 25–26.
41. Smythe, *Pershing,* 79.
42. Cooke, *Pershing and His Generals,* 101.
43. Frank P. Lahm, *The World War I Diary of Col. Frank P. Lahm, Air Service, A.E.F.* (Maxwell Air Force Base, Ala.: Historical Research Division, Air University, 1970), 80. The 2 June event and aftermath are detailed in Hudson, *Hostile Skies,* 57–58.
44. William Mitchell, letter to Benjamin Foulois, July 19, 1918, Mitchell Papers, Special Collections, LC.
45. Cited in Coffman, *The War to End All Wars,* 207, 369.
46. Kennett, *The First Air War,* 221. For an excellent discussion of the military value of aviation during World War I, see chap. 13.
47. Patrick Correspondence, File T-Z, Entry 228, Box 8, RG 18, NA.
48. Mortensen, "Air Service in the Great War," 48.
49. Gorrell, in a 29 Feb. 1924 letter (responding to a request from Patrick), presents a quintessential summary of the Bolling Mission and how various decisions were made regarding the production of aircraft, and for what reasons. See Gorrell to Patrick, 29 Feb. 1924, Nordyke and Marmon Co. letter no. 23, Entry 228, Box 1, RG 18, NA. Also see Hallion, *Rise of the Fighter Aircraft,* iii; and Patrick, *United States in the Air,* 11.
50. Patrick was in full agreement with the Bolling Commission recommendations that American manufacturers should emphasize DH-4 and Liberty engine production. See Patrick, *United States in the Air,* 12.

51. Mortensen, "Air Service in the Great War," 49.
52. Patrick, *United States in the Air,* 34–35.
53. Of course, Pershing had his critics. In fact, MMP committed such a criticism to his diary regarding Pershing's early reluctance to streamline his top-heavy headquarters. See Patrick diary entry for 22 Feb. 1918; also cited in Smythe, *Pershing,* 82. (Note: Smythe citation inadvertently notes 20 Feb. vice 22 Feb. for MMP diary entry.)

Chapter 4. Patrick and the Postwar AEF Air Service, 1918–1919

1. Patrick diary, entries for 13, 18, and 20 Oct. 1918.
2. Ibid., entry for 6–7 March 1919.
3. Ibid., entry for 13 Oct. 1919; Patrick, *United States in the Air,* 56.
4. Patrick, *United States in the Air,* 57.
5. Assistant Chief of Air Staff, Intelligence, Historical Division, *Organization of Military Aeronautics, 1907–1935,* 37, File 168.1-6B, AFHRA.
6. Patrick, letter to Pershing, "Proceedings of a Board on the Air Service," 19 May 1919, 6, File 167.404-5, AFHRA.
7. Futrell, *Ideas, Concepts, and Doctrine,* 28; Proceedings of Board on the Air Service, American Expeditionary Forces, 13 May 1919; Patrick, letter to Pershing, "Proceedings."
8. Patrick diary, entry for 8 June 1919.
9. See Davis, *Billy Mitchell Affair,* 57; and R. Earl McClendon, *Autonomy of the Air Arm* (Washington, D.C.: Air Force History and Museums Program, 1996), 36–37.
10. Patrick diary, entry for 31 May 1919.
11. Ibid., 21 June 1919.
12. Ibid., 27 June 1919.
13. Ibid., 6 May 1919.
14. Ibid., 27 June 1919.
15. Ibid., 1 July 1919.
16. Many, but not all, of Mrs. Bennett's letters to General Patrick were found in Patrick's personal correspondence file in RG 18, NA. His responses are located in the Bennett Collection.
17. Bream Cooley Patrick, the only child, graduated from West Point, saw service in World War II and Korea, and rose to the rank of colonel. He died in 1976.
18. Patrick diary, entry for 14 July 1919.
19. See "Gorrell's History of the American Expeditionary Forces Air Service, 1917–1919," RG 120, NA, or Maurer, ed., *U.S. Air Service in World War I.* It is to Patrick that we owe a debt of gratitude as he directed Gorrell to undertake the compilation of the lessons learned history of the Air Service, AEF, at the conclusion of the war.
20. See Patrick Correspondence.
21. Davis, *Billy Mitchell Affair,* 3–4.

22. War Department Telegram no. 2415-R, to Pershing, 7 Jan. 1919, paras. 8 and 17, Box 1, MS 17, USAFA. Also see Cooke, *U.S. Air Service in the Great War,* 211.
23. Patrick, *United States in the Air,* 69; Boyle, *Trenchard,* 309.
24. Maurer, ed., *U.S. Air Service in World War I,* 11; Frear Subcommittee Report, pt. 1, 66th Cong., 2nd sess., 1919, H. Rept. 637; Patrick, *United States in the Air,* 69.
25. MMP, letter to Mrs. Louis Bennett, 29 Oct. 1919, Bennett Collection. For a detailed look at the World War I demobilization process see Benedict Crowell and Robert F. Wilson, *Demobilization,* vol. 4 of *How America Went to War* (New Haven: Yale University Press 1921). For section dealing with Patrick and aviation surplus, see 212–13.
26. MMP, letter to Pershing, 11 Nov. 1919, Box 483, File 321.9, RG 18, NA.

Chapter 5. The First Round in the Postwar Fight for Air Service Independence, 1919–1921

1. John F. Shiner, "From Air Service to Air Corps: The Era of Billy Mitchell," in *Winged Shield, Winged Sword,* ed. Nalty, 1:72.
2. Cooke, *U.S. Air Service in the Great War,* 211.
3. MMP, letter to Mrs. Bennett, 10 Nov. 1919, 2, Bennett Collection.
4. Craven and Cate, *Plans and Early Operations,* 23–24.
5. Martha E. Layman, *Legislation Relating to the Air Corps Personnel and Training Programs, 1907–1939,* Army Air Forces Historical Studies no. 39 (Washington, D.C.: Army Air Forces Historical Office, 1945), 14, AFHRA.
6. Editorial, *Washington Star,* 13 Aug. 1919, editorial section, 1.
7. Mitchell Papers, April/May 1919, Manuscript Division, LC.
8. ACS, *Organization of Military Aeronautics,* 41.
9. McClendon, *Autonomy of the Air Arm,* 40.
10. ACS, *Organization of Military Aeronautics,* 42; McClendon, *Autonomy of the Air Arm,* 41; Maurer, ed., *U.S. Air Service in World War I,* 41.
11. 66th Cong., 2 Nov. 1919, 2:1571–1704.
12. Menoher to Pershing, "Request for Statement Regarding a Separate Air Service," 16 Dec. 1919, Box 482, File 321.9, RG 18, NA.
13. Ibid.
14. Pershing to Menoher, 12 Jan. 1920, Box 482, File 321.9F, RG 18, NA.
15. Bell to Menoher Board, "Comments Received from General Officers in August 1919, in Opposition to Creation of a Department of Aeronautics," File 168.6541-10, 1:1, AFHRA.
16. Ibid., 2.
17. Ibid., 3.
18. Shiner, "From Air Service to Air Corps," 73.
19. Maj. B. D. Foulois, "On the Necessity for the Creation of a Department of Aeronautics," statement made before the Committees on Military Affairs of the Senate and

House of Representatives, 66th Cong., 1st sess., 6 Oct. 1919, File 168.6541-10, 1:2, AFHRA; emphasis in original.
20. Shiner, *Foulois and the U.S. Army Air Corps,* 15.
21. Adjutant General's Office, *Army Register* (Washington, D.C.: GPO, 1 Jan. 1920), 14.
22. "Recommendations Concerning the Establishment of a Department of Aeronautics," prepared under the direction of Brig. Gen. Mitchell by Colonel Dodd, 17 April 1919, 4, in vol. 1 of the Jones Collection, MS 33, USAFA/SC.
23. See Davis, *Billy Mitchell Affair,* 57–58; and Jeffrey S. Underwood, *The Wings of Democracy: The Influence of Air Power on the Roosevelt Administration, 1933–1941* (College Station: Texas A&M Press, 1991), 9.
24. Secretary of War, annual report (1922), in AFHSO archives.
25. ACS, *Organization of Military Aeronautics,* 47; and Secretary of War, annual report (1921), 188.
26. William F. Trimble, *Admiral William A. Moffett: Architect of Naval Aviation* (Washington, D.C.: Smithsonian Institution Press, 1994), 72.
27. See Maurer, ed., *U.S. Air Service in World War I,* 43–52, passim; Davis, *Billy Mitchell Affair,* 60; Nalty, ed., *Winged Shield, Winged Sword,* 76.
28. Secretary of War, annual report (1919), 68–75.
29. Underwood, *Wings of Democracy,* 13.
30. See Maj. Gen. Charles T. Menoher, "The Future of the Air Service," *U.S. Air Service* 1, no. 3 (April 1919): 10–11 (all issues of *U.S. Air Service,* the official publication of the Army and Navy Air Service Association, can be found in File 167.63, AFHRA); and Brig. Gen. William Mitchell, "Air Leadership," and Sen. George E. Chamberlain, "Military Aeronautics: No General Staff Control of Army Air Service," both in *U.S. Air Service* 1, no. 4 (May 1919): 13–15, 16–17. While a majority of Air Service officers leaned toward independence, it was by no means a universal feeling, especially outside of military aviation circles. See Warren A. Trest, *Air Force Roles and Missions: A History* (Washington, D.C.: Air Force History and Museums Program, 1998), 30.
31. William Mitchell, *Our Air Force: The Keystone of National Defense* (New York: E. P. Dutton, 1921), xvii–xviii, 199–216.
32. President's Aircraft Board, *Hearings Before the President's Aircraft Board* (Washington, D.C.: GPO, 1925), 497.
33. Davis, *Billy Mitchell Affair,* 66–67.
34. File 3084-45, Old Navy Records, RG 72, NA. Although this incident is mentioned in Burke Davis's 1967 book, *The Billy Mitchell Affair* (89), it is curiously absent from William F. Trimble's 1994 biography, *Admiral William A. Moffett.*
35. Davis, *Billy Mitchell Affair,* 90.
36. Mitchell Papers, 201 File, Part 1, Manuscript Division, LC. This episode is treated in detail in Burke Davis's book, relying principally on the Library of Congress's Mitchell Papers.

37. *New York Times,* 2 Aug. 1921, sec. 2, 1.
38. *Cong. Rec.,* 67th Cong., 1st sess., 11 April 1921–23 Nov. 1921, 500. Also cited in James Philip Tate, *The Army and Its Air Corps of Army Policy Towards Aviation, 1919–1941* (Maxwell Air Force Base, Ala.: Air University Press, 1998), 24. Borah and Mitchell shared a like mind on the potential economy and defensive aspects of an air fleet. While at Langley preparing for the bombing tests, Mitchell periodically sent fresh oysters to Borah via planes headed up to Bolling Field outside of D.C. See Davis, *Billy Mitchell Affair,* 99.
39. Tate, *The Army and Its Air Corps,* 25. Tate's recently published 1976 dissertation (Air University Press, 1998) provides insightful information on this Air Service/Air Corps era.
40. 201 File, part 1, Mitchell Papers, LC.
41. Davis, *Billy Mitchell Affair,* 114.
42. Joint Board report can be found in the *Cong. Rec.,* 67th Cong., 1st sess., 11 April 1921–23 Nov. 1921, 8625–26; also see Tate, *The Army and Its Air Corps,* 25.
43. U.S. Naval Institute *Proceedings* 47 (Nov. 1921): 1828–29.
44. Tate, *The Army and Its Air Corps,* 25.
45. *New York Times,* 14 Sept. 1921, sec. 1, 2.
46. Arnold, *Global Mission,* 106. Also see Tate, *The Army and Its Air Corps,* 26.
47. There are various historical accounts of this showdown and why Weeks acted as he did. See Levine, *Mitchell,* 270; Hurley, *Billy Mitchell,* 69; Futrell, *Ideas, Concepts, and Doctrine,* 37; Arnold, *Global Mission,* 104–5. Also see Tate, *The Army and Its Air Corps,* 26, and Bingle, "Building the Foundation," 40 n. 29.
48. Adjutant General, letter to CoAS, 25 Aug. 1921, File 167.411-5-1, AFHRA.
49. Patrick, *United States in the Air,* 85–86.
50. Ibid., 73.
51. Maj. Gen. H. F. Hodges, letter to MMP, 22 Sept. 1921, Entry 228/229, Box 4, RG 18, NA.
52. See Entry 228/229, Boxes 4 and 5, RG 18, NA.
53. Drum, letter to Patrick, 22 Sept. 1921, Entry 228, Box 4, RG 18, NA. The letter read, in part, "I am indeed happy that this deserved and just reward has come to you. No one knows more than I do the fine work that you did in the A.E.F and no one has felt more than I have the fact that you have received very little credit therefor."
54. DeWitt S. Copp, *A Few Great Captains* (McLean, Va.: EPM Publications, 1980), 64. Congratulatory letters from the Army's general officer corps were numerous and invariably mentioned, in one form or another, the hope that, "your [appointment] signifies a period of increased efficiency of the Air Service." See, for example, Maj. Gen. C. P. Summerall, letter to MMP, 23 Sept. 1921, Entry 228, Folder N-S, Box 8, RG 18, NA.
55. Davis, *Billy Mitchell Affair,* 120.
56. Arnold, *Global Mission,* 90.

Notes to Pages 58–62

Chapter 6. "To Command in Fact as Well as in Name"

1. Patrick, *United States in the Air,* 69–70; MMP, letter to Louis Bennett, 29 Oct. 1919, 2–3, Bennett Collection.
2. MMP, letter to Louis Bennett, 10 Nov. 1919, 2–3, Bennett Collection.
3. MMP, letter to Louis Bennett, 20 Dec. 1919, Bennett Collection.
4. Patrick, *United States in the Air,* 70.
5. For an excellent overview of Billy Mitchell's "publicist" character, see Shiner, "From Air Service to Air Corps," 71–100; also see Robin Higham, *Air Power: A Concise History* (Manhattan, Kan.: Sunflower University Press, 1988), 33.
6. Patrick, *United States in the Air,* 83.
7. War Department Special Orders No. 231-0, 5 Oct. 1921, Series 228, Box 6, File M-P, RG 18, NA.
8. Patrick, *United States in the Air,* 73.
9. Ibid., 85–86.
10. This confrontational episode between Patrick and Mitchell is detailed in Patrick's *United States in the Air,* 86–89.
11. Arnold, *Global Mission,* 106.
12. Patrick, *United States in the Air,* 88. Contrary to Patrick and Harbord's version of events, Isaac Don Levine in his *Mitchell* (273) claims that Mitchell alone was responsible for insisting on a written statement of his responsibilities, with the Mitchell-inspired memorandum being finalized and dated 18 October. The final memorandum (in Mitchell Papers, LC) is dated 18 October, the majority of which was produced by Patrick prior to Monday's meeting with Harbord. The revision incorporated some minor changes made jointly by Patrick and Mitchell at the meeting's conclusion. Davis, in citing this incident in *The Billy Mitchell Affair,* characterizes the Patrick/Mitchell relationship thus: "no one routed [Mitchell] so completely in a personal clash as Mason Patrick did." See p. 120.
13. File 321.9, Box 402, RG 18, NA.
14. "Mitchell and Mason Patrick Seek a *Modus Vivendi,*" para. 2, Box 11, Card 2.30, MS 33, USAFA/SC.
15. Patrick, *United States in the Air,* 89.
16. Maj. Gen. Francis J. Kernan, letter to MMP, 28 Sept. 1921, Entry 228/229, Box 5, RG 18, NA.
17. MMP, letter to Maj. Gen. Francis J. Kernan, 2 Nov. 1921, in ibid.
18. Adjutant General Special Order, 23 Nov. 1921, Mitchell's 201 File, Special Collections, LC. For complete details of Mitchell's European tour, see "Report of Inspection Trip to France, Italy, Germany, Holland, and England, Made During Winter of 1921–1922," File 167.404-13, AFHSO.
19. Patrick, letter to Mitchell, 7 Dec. 1921, Mitchell's 201 File, LC.

20. Davis, *Billy Mitchell Affair*, 134.
21. For Mitchell's report, see, "Report of Inspection Trip to France, Italy, Germany, Holland, and England, Made during the Winter of 1921–1922 by Brigadier General Mitchell, First Lieutenant Clayton Bissell and Aeronautical Engineer Alfred Verville," 1922, Air Service Records, Microfiche File 167.404, Roll A1538, Frames 1–122, USAF History Support Office, Washington, D.C.
22. Prior to Mitchell's court-martial in 1925, his defense requested a listing all formal suggestions/proposals submitted by Mitchell bearing on the improvement of the Air Service and its mission. The total came to 127; these being limited to proposals that were forwarded to the War Department. From 1919 through 1925, at the Air Service headquarters level, over twice this total were returned disapproved. See "Court Martial," Entry 166.250, Box 382, Files 4–6, RG 18, NA.
23. Patrick, *United States in the Air*, 91–92; Davis, *Billy Mitchell Affair*, 143.
24. CoAS annual report (1921), contained within secretary of war's annual report (1921), 183–94.
25. Ibid., sec. 1, 1.
26. While early in his tenure as Air Service chief Patrick may have looked on the positive side of the 1920 act, in 1923 he described it as "unbalanced and entirely inadequate." See CoAS annual report, 1923 (13 Sept. 1923), 48, File 167.4011, AFHRA.
27. CoAS annual report (1921), 183.
28. At the time, the secretary of war ignored Patrick's recommendation that this issue be addressed, but Patrick later revived the issue when he drafted the legislation for what became the 1926 Air Corps Act. See ACS, *Organization of Military Aeronautics*, 59.
29. CoAS annual report (1921), 185.
30. Ibid.
31. For a discussion of doctrine, roles, missions, reorganization, and Patrick's role therein, see Bingle's thesis, "Building the Foundation," chap. 3, "Developmental Politics, 1921–1925," and Futrell, *Ideas, Concepts, and Doctrine*, 39–53.
32. Futrell, *Ideas, Concepts, and Doctrine*, 40–41; Bingle, "Building the Foundation," 44–45.
33. One can make a strong case that all World War I air-related activity was more tactics and techniques than doctrine, especially in the Army Air Service. But tactics and technique often form the basis for doctrine. Mitchell himself considered his St. Mihiel set piece attack as the ideal use of air power until his return from his European tour in 1922.
34. Patrick, *United States in the Air*, 90.
35. See I. B. Holley Jr., "An Enduring Challenge: The Problem of Air Force Doctrine," *The Harmon Memorial Lectures in Military History*, no. 16 (Colorado Springs: U.S. Air Force Academy, 1974), 3.

36. Office of the Air Service Chief, memorandum, "Organization of the Office of the Chief of Air Service," 1 Dec. 1921, Organization, Miscellaneous, Box 489, File 321.9, RG 18, NA; Futrell, *Ideas, Concepts, and Doctrine,* 40; U.S. Air Force, *The Development of Air Doctrine in the Army Air Arm, 1917–1941,* USAF Historical Studies no. 89 (Maxwell Air Force Base, Ala.: Historical Division, Air University, 1955), 16; Bingle, "Building the Foundation," 43.
37. "Organization of the Office of the Chief of Air Service," 21 Dec. 1921, 9.
38. The Air Service Field Officers School was redesignated the Air Service Tactical School in 1922 and the Air Corps Tactical School in 1926, which subsequently moved from Langley to Maxwell Field, Montgomery, Alabama, on 1 July 1931. See Robert T. Finney, *History of the Air Corps Tactical School, 1920–1940* (Washington, D.C.: OAFH, 1994), passim.
39. In writing this doctrinal treatise, Major Sherman was most probably assisted by the very capable commander of the Air Service Officers School, Maj. Tommy Milling. The basis for Sherman's work was a June 1919 document entitled "Manual for Air Service Operations," published under Col. Edgar S. Gorrell's name while he was ACS, Air Service, AEF, but in all probability it was written by Sherman, who was on Gorrell's staff. See Futrell, *Ideas, Concepts, and Doctrine,* 1:40–41; John D. Parker, *The Early Development of United States Air Doctrine,* Professional Study no. 6024 (Maxwell Air Force Base, Ala.: Air War College, 1976), 81–84; and Flugel, "United States Air Power Doctrine," chap. 4.
40. Bingle, "Building the Foundation," 44–45; Futrell, *Ideas, Concepts, and Doctrine,* 40–41.
41. Mitchell, memo to Director of Air Service, 16 April 1919, as cited in Flugel, "United States Air Power Doctrine," 55. For a detailed discussion of the roles of Mitchell, Sherman, and Gorrell concerning the early years of doctrinal development, see Parker, *Early Development of United States Air Doctrine.*
42. CoAS annual report (1921), 185.
43. See "Notes from the Air," *U.S. Air Service,* Nov. 1921 through January 1927. This magazine section, among other topics, noted the official travels of the respective Air Service chiefs.
44. Mason M. Patrick, "The Air Service," lecture at the Army War College (hereafter Air Service lecture), 25 April 1922, 7, Curriculum File, 33-117/a, Army War College, MHI.
45. Air Service lecture, 2, 9, 10; and CoAS annual report (1921), 185.
46. CoAS annual report (1921), 185.
47. Futrell, *Ideas, Concepts, and Doctrine,* 74–77.
48. Mason M. Patrick, "The Use of Aircraft in War," *American Unity Forward* (Oct. 1922).
49. Patrick, *United States in the Air,* 79–81.
50. MMP, Office Diary, 7 Oct. 1921, Series Box 7, Folder 1, Foulois Papers.

51. Charles Chatfield, *The American Peace Movement: Ideals and Activism* (New York: Twayne Publishers, 1991), chap. 3; Russell F. Weigley, *History of the United States Army* (New York: Macmillan, 1967), 402; Bingle, "Building the Foundation," 55, 89 n. 27.
52. In an inaugural address that emphasized "economy" and "tax reduction" as its two main points (and was characterized as "vague" by more than one newspaper), Coolidge stated, "I favor the policy of economy, not because I wish to save money, but because I wish to save people." See *The Independent* 114, no. 3902 (14 March 1925): 280.
53. *The Independent* 114, no. 3902 (14 March 1925): 280.
54. Edwin H. Rutkowski, *The Politics of Military Aviation Procurement, 1926–1934* (Columbus: Ohio State University Press, 1966), 55. Although this book's focus is a discussion concerning the implementation of the Air Corps Act of 1926 (Patrick's Legislation) Patrick is not mentioned, even though he did not retire until eighteen months after the act's passage.
55. Bingle, "Building the Foundation," 57.
56. Air Service lecture, 7.
57. Maj. William C. Sherman, "Air Tactics," 1921, sec. 2, 1.
58. Patrick, *United States in the Air,* 95–97; U.S. Congress, House Committee on Appropriations, *Hearings on the War Department Appropriation Bill for 1925,* 68th Cong., 1st sess., 1924, 901–2, as cited in Bingle, "Building the Foundation," 66, 90 n. 52.
59. See chap. 6, "Fundamental Conceptions," the basis for Training Regulation 440-15; Futrell, *Ideas, Concepts, and Doctrine,* 40–41, 50; Bingle, "Building the Foundation," 44.
60. "Minutes of Monday Meeting of Division Chiefs," MS 17 (Reports), Box 7, Series 4, Folder 5, 12, USAFA/SC.
61. Patrick, *United States in the Air,* 99–100.
62. Davis, *Billy Mitchell Affair,* 133–35.
63. Ibid., 143; Patrick, *United States in the Air,* 79–81.
64. Tate, *The Army and Its Air Corps,* 31.
65. Harry H. Ransom, "The Air Corps Act of 1926: A Study in the Legislative Process," Ph.D. diss., Princeton University, 1953, 124.
66. Tate, *The Army and Its Air Corps,* 30.
67. See "Speech Delivered by Major General Mason M. Patrick at the American Defense Society, N.Y.C., on January 15, 1923," Entry 228/229, Box 1, RG 18, NA.
68. The World War I–era DH-4 aircraft that the Air Service employed were actually heavily modified and updated to the DH-4B model in the 1920–24 time frame.
69. Secretary of War, annual report (1922), 261–63.
70. House, *Inquiry into Operations of the United States Air Service; Hearings Before the*

Select Committee of Inquiry into Operations of the United States Air Service, 68th Cong., 2nd sess., 1924, 519.
71. Futrell, *Ideas, Concepts, and Doctrine,* 1:40.

Chapter 7. The Lassiter Board and "Fundamental Conceptions"

1. See Maj. Gen. Mason M. Patrick, "Recommendations for Improving the Air Service," *U.S. Air Service* (Feb. 1923): 15–16.
2. See, for example, "Address of Maj. Gen. M. M. Patrick at Lu Lu Temple," Philadelphia, 12 Feb. 1923; "Speech delivered by General Patrick at Akron, Ohio, 7 November 1923"; "Address delivered by General Mason M. Patrick before Chamber of Commerce, Harrisburg, Penna., 6 October 1923"; "Address delivered by Gen. Patrick on Dec. 11 1923 before Association of Life Insurance Presidents, Chicago, Illinois." All in Entry 228, Box 1, RG 18, NA.
3. See Listing of Speeches and Articles, Entry 228, Box 1, RG 18, NA.
4. See, for example, Maj. Gen. Mason M. Patrick, "The Development of Air Traffic in the Next Few Years," *New York Times,* 13 March 1923.
5. "Speech Delivered by Major General Mason M. Patrick at the American Defense Society, N.Y.C., on January 15, 1923," Entry 228/229, Box 1, RG 18, NA. As was routine, Patrick's staff compiled related editorials concerning the speech for Patrick's review. Although the major New York and Washington newspapers gave very positive coverage to the speech, Patrick went out of his way to send a personal letter to the editor of the *Troy* (N.Y.) *Times,* thanking him for their support. See MMP, letter to editor, *Troy Times,* 20 Jan. 1923, Series 228, Box 5, RG 18, NA.
6. "Air Force Tactics," speech delivered at the Army War College, Nov. 1923, 5, Entry 228/229, Box 1, RG 18, NA. This speech, complete with slides (or "plates") included Patrick's diagrammatic vision of an offensive "Air Force" that would operate "*entirely independent*" of ground troops," 7.
7. At that time, Mitchell was on an inspection tour in the Midwest combined with a trip to Detroit to see his soon-to-be second wife, Betty Miller.
8. "Minutes of Monday Meeting of Division Chiefs."
9. See, for example, Maj. W. H. Frank, letter to Maj. Gen. Mason M. Patrick, c/o Maj. H. H. Arnold, Rockwell Air Intermediate Depot, Coronado, Calif., 10 April 1923, Series 228, Box 6, File M-P, RG 18, NA.
10. Davis, *Billy Mitchell Affair,* 128–30, 145–46.
11. Patrick, Office Diary, 1922–26, Foulois Papers, MS 17, Box 6, Series 3, Folder 9, USAFA/SC.
12. "Minutes of Monday Meeting of Division Chiefs," passim.
13. Gen. Ira C. Eaker, interview by Lt. Col. Joe B. Green, Washington, D.C., 28 Jan. 1972, p. 30, on file at Army War College, Carlisle Barracks, Pa.
14. "Speech delivered by General Patrick at Annual Banquet of Aero Club of America,

given in New York City on January 9, 1922," 3, Entry 228, Box 1, File A-D, RG 18, NA.
15. Trest, *Air Force Roles and Missions,* 39.
16. See letters from Wainwright to Patrick and responses, 19 Dec. 1921, 5 Oct. 1922, and 18 Dec. 1922, Series 228, Box 8, File W-Z, RG 18, NA; also *New York Times,* 19 May 1923, 2.
17. See Patrick correspondence, and Patrick diary.
18. See, for example, Aero Club speech; "Pilots in Regular Army Air Service who would be capable of taking the field as pilots with one week's training," Series 228, Box 6, File M-P, RG 18, NA; Patrick, *United States in the Air,* 188–89.
19. Tate, *The Army and Its Air Corps,* 19.
20. Maj. Gen. H. H. Tebbetts, the Adjutant General, War Department, letter to CoAS, "Preparation of a Project for the Peace Establishment of the Air Service," 18 Dec. 1922, File 145.93-101, AFHRA (hereafter "Preparation of a Project").
21. First endorsement, Maj. Gen. Mason M. Patrick to The Adjutant General, 19 Jan. 1923, File 145.93-101, AFHRA.
22. "Preparation of a Project," para. 3.
23. Third endorsement, Maj. Gen. Mason M. Patrick to Adjutant General (hereafter Patrick endorsement to AG), 7 Feb. 1923, File 145.93-101, 2–3, AFHRA.
24. Patrick, "Recommendations for Improving the Air Service," 15–16.
25. Patrick endorsement to AG.
26. Ibid.
27. Patrick endorsement to AG; Bingle, "Building the Foundation," 46; Futrell, *Ideas, Concepts, and Doctrine,* 42.
28. Patrick endorsement to AG.
29. "War Department Committee Report on the Organization of the Air Service," File 145.93-101, AFHRA (hereafter Lassiter Board). Only five members were on the General Staff proper, while two (Brigadier General Drum and Colonel Hunt, QMC) were detailed to it.
30. For Dickman Board Report, see File 167.404-6, AFHRA. Although Lassiter spoke unabashedly of his desire to keep air elements integral to each division (during both the Dickman and Lassiter Boards), he did so with the greatest of praise for the airmen's accomplishments and capabilities.
31. "Report of a Committee of Officers," 2, File 145.93-101, AFHRA; Maurer, *Aviation in the U.S. Army,* 541 n. 13.
32. The complete notes of the Lassiter Board proceedings can be found in Files 145.93-101/102 at AFHRA, complete with a memorandum from General Drum directing that the notes be saved, "as they may prove of value some day." They ultimately did, being used as the basis for action by the General Staff when the Morrow Board again looked at the Air Service issue in late 1925.

33. See "Comments Received from General Officers in August 1919, In Opposition to Creation of A Department of Aeronautics," File 168.6541-10, Vol. 1, AFHRA.
34. Lassiter Board, "Minutes," 22 March 1923, 2, and 23 March 1923, 8, 11.
35. Bingle, "Building the Foundation," 47.
36. Excerpt taken from "Report of the Special Committee Appointed by the Director, War Plans Division, to Define the General Plan of Organization to be Adopted for the Army of the United States, Provided by the Act of June 4, 1920," which was quoted in "Report of a Committee of Officers Appointed by the Secretary of War," dated 27 March 1923, 3 (hereafter referred to as Report of the Special Committee). This document was the finalized version signed by Secretary Weeks on 24 April 1923. Located in File 167.404-8, AFHRA.
37. Lassiter Board, "Minutes," 23 March 1923, 11.
38. During one part of the division observation discussion, General Drum stated: "I feel so strongly on the question that I went around and talked with some men that operated in France close to the front line, in order to see whether our views were wrong or not. [One] man, Colonel [George C.] Marshall, I would like to have come before the board. His opinion would carry a good deal of weight on the question of aviation with the division. Lassiter Board, "Minutes," 22 March 1923, 5. (The minutes did not list how individual board members voted on the issues.)
39. Lassiter Board, "Minutes," 23 March 1923, 9.
40. Report of the Special Committee, 4.
41. Lassiter Board, "Minutes," 23 March 1923, 10.
42. Maj. Gen. Mason M. Patrick, notes on Lassiter Committee, "Jones" Papers, File 168.6501-40, AFHRA (hereafter Notes on Lassiter). This issue was referred to the Joint Board, ultimately without successful resolution.
43. Ibid., 5.
44. Ibid., 6.
45. "Report of a Committee of Officers," 6.
46. Patrick to Wainwright, "Memorandum for the Assistant Secretary of War," 5 April 1923, Entry 189, Box 6, RG 18, NA.
47. Ibid., Enclosure, 1.
48. Ibid., Enclosure, 3.
49. U.S. War Department, *Field Service Regulations, United States Army, 1923* (Washington, D.C.: GPO, 1924), 11, quoted in Futrell, *Ideas, Concepts, and Doctrine,* 43.
50. U.S. War Department, *Field Service Regulations,* 21–24, quoted in Futrell, *Ideas, Concepts, and Doctrine,* 43; Bingle, "Building the Foundation," 49.
51. Patrick diary.
52. MMP, letter to Pershing, 11 Nov. 1919, File 321.9, Box 483, RG 18, NA; and Patrick, *United States in the Air,* 101–2.
53. James Tate, in *The Army and Its Air Corps,* states: "After studying Patrick's proposals,

the Lassiter Board gave full endorsement" (19), and also states: "The most important aspect of the Lassiter Board report was its acceptance of General Patrick's plan to divide the air arm according to tasks. The observation air arm would be an integral part of divisions, corps, and armies, with a reserve under general headquarters" (20). On the contrary, the board did not fully endorse General Patrick's plan. Patrick was for eliminating observation assets from division level. See Lassiter Board, "Minutes," 22 March 1923, 4.

54. "Fundamental Conception of the Air Service: Prepared Under the Direction of the Chief of Air Service, 1923," File 167.404-10, AFHRA.
55. During 1922 Capt. William S. Sherman's manuscript on air tactics was revised into manual form. This was used as the basis for "Fundamental Conceptions." See Futrell, *Ideas, Concepts, and Doctrine,* 41.
56. "Mason Patrick's Toupee" (Murray Green Collection), Box 4, Card 2.23, Arnold Papers, USAFA/SC.
57. "Fundamental Conceptions," 8.
58. Ibid., 9.
59. As noted in Rebecca Hancock Cameron, *Training to Fly: Military Flight Training, 1907–1945* (Washington, D.C.: Air Force History and Museums Program, 1999), 219–20.
60. Levine, *Mitchell,* 282–83.
61. Patrick, letter to Lt. Col. R. C. Kirtland, 4 Dec. 1923, Entry 228/229, Box 5, RG 18, NA. Also cited in Bingle, "Building the Foundation," 49.
62. J. Mayhew Wainwright, "The Situation in the Air Service," *U.S. Air Service* (April 1923): 19.
63. See *Washington Star, New York Times, U.S. Air Service,* and *The Army and Navy Journal,* Oct. 1922–March 1923.

Chapter 8. Patrick Takes on a Pair of Wings, the Navy, and the Army General Service Schools

1. H. A. Dargue, "Training a Major-General to Fly," *U.S. Air Service* (Sept. 1923): 18–22; Bingle, "Building the Foundation," 16.
2. Dargue, "Training a Major-General to Fly," 22.
3. See Patrick correspondence.
4. Patrick, *United States in the Air,* 112–13. Patrick's blue and yellow DH-4B, Tail Number A.S. 63786, is on permanent display at the U.S. Air Force Museum, Wright-Patterson AFB, Dayton, Ohio.
5. Maj. Gen. Mason M. Patrick, Confidential Memorandum, "Bombing Experiments," 19 March 1923, 1, Series 228, Box 2, RG 18, NA.
6. "Office Diary of Maj. Gen. Mason M. Patrick, Chief of Air Service," 1922–24,

passim, MS 17, Box 6, Series 3, Folder 9, USAFA/SC; also Davis, *Billy Mitchell Affair,* 151.
7. The principal results of the Lassiter Board appeared in the press on 19 Oct. 1923.
8. Denby, letter to Weeks, 20 Feb. 1923, Series 228, Box 2, RG 18, NA.
9. Patrick, "Bombing Experiments," 1, para. 1.
10. Ibid., 1, para. 4.
11. Ibid., para. 5.
12. But Patrick was busy himself with an extended inspection tour of Air Service squadrons and operations during part of this time (April through early May).
13. Patrick, memorandum to Pershing, 30 June 1923, Box 62, RG 94, NA.
14. Levine, *Mitchell,* 287; Davis, *Billy Mitchell Affair,* 151.
15. Levine, *Mitchell,* 292–93.
16. *Air Service News Letter,* 23 Aug. 1923, 8, cited in Maurer, *Aviation in the U.S. Army,* 124.
17. In a November 1923 speech, just over two months after the *Virginia* and *New Jersey* were dispatched, Patrick stated: "The object of these maneuvers was to show the mobility of aviation . . . and give the personnel some needed practice in bombardment and to let them see a battleship sunk under air attack. No officer who ever witnessed a ship sunk by aircraft would ever hesitate in pushing an attack against them in time of war." Speech, "Air Force Tactics," 5, Entry 229, Box 5, RG 18, NA.
18. Maj. Gen. Mason M. Patrick, "The Engineering Problems of Aviation," lecture before the Harvard Engineering Society, Dec. 1923, passim, Series 228, Box 5, RG 18, NA; and Davis, *Billy Mitchell Affair,* 152–53.
19. *Time,* 9 July 1923.
20. Davis, *Billy Mitchell Affair,* 153.
21. Ibid., 153–54.
22. Maurer, *Aviation in the U.S. Army,* 127.
23. Departing on 23 Oct. 1923, Mitchell and his wife visited Hawaii, the Philippines, China, India, and Japan. (The India leg was to be at Mitchell's expense.) Mitchell returned to Washington in July 1924. See Levine, *Mitchell,* chap. 10, and particularly Davis, *Billy Mitchell Affair,* chapters 12–14, an especially well-researched and documented examination of this Pacific excursion.
24. *Army and Navy Journal* 61, all issues of Sept. and Oct. 1923. Also cited in Bingle, "Building the Foundation," 50.
25. *Army and Naval Journal* 61 (6 Oct. 1923): 120; Bingle, "Building the Foundation," 49–50.
26. See "Testimony of William Mitchell," House of Representatives, Committee on Military Affairs, Washington, Friday, February 5, 1926, Air Service Records, Microfiche File 167.404-13, Roll A1538, Frames 1003–1004, USAF History Support Office, Washington, D.C.

27. For numerous examples see Entry 228, Boxes 1–8, RG 18, NA.
28. Engineering Problems lecture, and attachment, "Synopsis," Dec. 1923, Series 228, Box 5, RG 18, NA.
29. For example: speech, "National Conference on Education for Highway Engineering," 26 Oct. 1922, Entry 229, Box 5, RG 18, NA; and numerous others in Boxes 1–8.
30. Memorandum, 12 Dec. 1921, "War College Lecture Data," and attachment: "Suggested Speech for Major General Mason M. Patrick Before the Army War College," Washington, D.C., March 1922, Entry 228, Box 4, RG 18, NA.
31. Air Service lecture, 7.
32. Maj. Gen. H. E. Ely, letter to Maj. Gen. Mason M. Patrick, 5 Oct. 1923, Entry 228, Box 4, RG 18, NA.
33. Ibid.
34. Army War College lectures on 27 Nov. 1923: "The Development of Military Aeronautics in the United States as Compared with that in France, Great Britain, Japan, Italy and Russia," and "Air Force Tactics," Entry 229, Box 5, RG 18, NA.
35. See chap. 7 for a discussion on Patrick's doctrinal metamorphosis.
36. Bingle, "Building the Foundation," 49.
37. "Fort Leavenworth Lecture." Two copies in file; cited is the one with marginalia reading: "Revised: As Delivered by Gen. Patrick," 27 March 1924, Entry 228, Box 1, RG 18, NA.
38. Ibid., 1.
39. Ibid., 6.
40. Ibid., 8, 10.
41. Ibid., 12. What Patrick described in 1924 is the modern-day Aerospace Expeditionary Force, or AEF. To this writer's knowledge, this unique historical parallel has not been hitherto recognized.
42. "Air Force Tactics," 1–2.
43. Letters, Arnold to Patrick and replies, Nov. 1922–May 1924, Entry 228, Box 1, RG 18, NA; Copp, *A Few Great Captains,* 34; J. E. Kuhn, letter to Patrick, 31 Jan. 1926, Entry 228, Box 3, RG 18, NA.
44. Bingle, "Building the Foundation," 54–55.

Chapter 9. Patrick's Search for Economy and Efficiency

1. CoAS annual report (1922), 27–28.
2. Patrick, *United States in the Air,* 11–101.
3. Maj. L. W. McIntosh, letter to Maj. Gen. Mason M. Patrick, May 27, 1924, Entry 228/229, Box 5, RG 18, NA.
4. See extensive correspondence in the Mitchell Papers, Manuscript Division, LoC.
5. See above, chap. 4, pp. 42–43.
6. While he was in Europe, one of Patrick's last acts prior to his return to the States, on 7

June 1919, was to formally request that the French minister of war send 2,260 foreign aircraft to the United States as part of a 55-million-franc liquidation settlement. Also, Germany had to surrender a number of aircraft. This, combined with the surge in U.S. production in the last months of the war, ultimately resulted in 3,117 "serviceable" planes when Patrick again took over the Air Service. See CoAS annual report (1921), 29; and "Materiel Organization," chap. 7, pp. 3–4, of "Early History of the Air Service," unpublished manuscript, n.d., AFHSO.

7. CoAS annual reports (1921), 21, and (1922), 29.
8. CoAS annual report (1922), 20.
9. "Statistics on Fatalities in Aviation: United States Army Air Service," 1, File 0.32A, Box 52, RG 18, NA.
10. U.S. Congress, House, Committee on Appropriations, *Hearings on the War Department Appropriations Bill for 1924,* 67th Cong., 2nd sess., 1923, 495.
11. "Statistics on Fatalities in Aviation," 1.
12. Hallion, *Rise of the Fighter Aircraft,* 28; Gorrell to Patrick, 29 Feb. 1924, Nordyke and Marmon Co. letter no. 23, 4, Entry 228, Box 4, RG 18, NA.
13. Bilstein, *American Aerospace Industry,* chaps. 1–2, passim; and Rutkowski, *Politics of Military Aviation Procurement,* chap. 2.
14. "Early History of the Air Service," chap. 7, 1–2.
15. Patrick, *United States in the Air,* 101–2.
16. Within the first three weeks after Patrick took over as Air Service chief, courtesy calls were paid by the president of the Manufacturers Aircraft Association, and the presidents (and one general counsel) of five aircraft manufacturing firms (Aero Transport, Martin, L.W.F., Curtiss, and Loening). Patrick inspected McCook Field shortly thereafter: 24–25 Oct. See entries, 10–17 Oct., Office Diary of MMP, 1921, 5–20, 24, Foulois Papers, Series Box 7, Folder 1, USAFA/SC; and CoAS Correspondence, 1921–27, Entry 228/229, Boxes 1–8, RG 18, NA.
17. The general descriptions by individuals who knew and served with Patrick are consistently positive, except for one family: the H. H. "Hap" Arnolds. Gen. Ira Eaker, who served on Patrick's staff, described him as "very intelligent" and "dynamic"; Maj. Gen. Robert L. Walsh, who replaced Hap Arnold (when Patrick exiled Arnold to Fort Riley in 1926), stated, "Patrick was wonderful." Joseph Litchfield, president of Goodyear Corporation, after an office call with Patrick: "You know, I have never been more beautifully handled. I didn't realize we had people in government who were of that kind, and who are as smart." See MS 33, "Hap Arnold" (Murray Green Collection), Box 4, Cards 2.23, 2.32, 2.45, USAFA/SC.
18. Patrick, *United States in the Air,* 102.
19. Patrick office diary, entry for 7 Oct. 1921, 1.
20. CoAS annual report (1921), 15.

21. Memorandum initialed by Patrick, 3 Feb. 1923, Entry 228, Box 4, RG 18, NA.
22. Memorandum, "In Considering Industrial Preparedness for War a Study Must Be Made of the Aeronautical Industry," n.d. (but after 20 Feb. 1923, based on marginalia), Entry 228, Box 4, RG 18, NA.
23. Ibid., 2.
24. *New York Times,* 19 Oct. 1923, 7; Report of the Special Committee, 5.
25. Speech: "Address by Colonel Dwight F. Davis, Assistant Secretary of War, Before the St. Louis Aeronautic Corporation and the Flying Club of St. Louis, at St. Louis, October 1, 1923," Entry 228, Box 1, RG 18, NA.
26. See "Aviation in the United States During 1923," *L'Air,* Nov. 1923, in Entry 229, Box 4, RG 18, NA.
27. Patrick, letter to McIntosh, 26 Nov. 1923, Entry 228, Box 5, RG 18, NA.
28. McIntosh, confidential memorandum to Patrick, 27 May 1924, Entry 228, Box 5, RG 18, NA.
29. "Materiel Organization," 4. Patrick, an engineer himself, developed a good rapport with Major Curry who stayed in that position until 1927. Patrick and Curry corresponded on a range of technical topics. See Entry 228, Box 4, RG 18, NA.
30. These two incidents described in Davis, *Billy Mitchell Affair,* 142. No specific source is cited by Davis, save for "recollections of several of Mitchell's contemporaries."
31. Patrick, memorandum to Mr. R. W. Ireland, 5 Jan. 1925, Entry 228, Box 4, RG 18, NA.
32. Patrick's testimony before U.S. Congress, House of Representatives, *Report of the Select Committee of Inquiry into Operations of the United States Air Services, House Report No. 1653* (hereafter Lampert Report) (Washington, D.C.: GPO, 1925), 130. Key extracts also found in File 167.404-7, AFHRA.
33. *Republican Campaign Textbook, 1924,* as cited in Tate, *The Army and Its Air Corps,* 29.
34. "Materiel Organization," 5; "Annual Report of the Chief of the Air Corps, 1928," 70; Bingle, "Building the Foundation," 99.
35. *Material Research and Development in the Army Air Arm, 1914–1945,* Army Air Force Historical Studies no. 50 (Army Air Forces Historical Office, Nov. 1946), 22, File 101-50, AFHRA.
36. Patrick, *United States in the Air,* 103.
37. President's Aircraft Board, *Hearings Before the President's Aircraft Board* (Washington, D.C.: GPO, 1925), 65; Also see "Testimony of Gen. M. M. Patrick, Maj. W. J. Kilner, Maj. T. G. Lamphier and others before the Morrow Board, Sept. 21-Oct. 13, 1925," File 248.211-61V, AFHRA.
38. Tate, in *The Army and Its Air Corps,* in describing Patrick's frame of mind in 1924 states: "The airplane shortage and the fact that many aircraft were obsolete also was, in part, a result of Patrick's decision to emphasize research and development of new

aircraft rather than standardization and procurement of designs that would have been obsolete before they could have been put into service" (33–34).

39. CoAS annual reports (1921–31); Rutkowski, *Politics of Military Aviation Procurement,* 256; Bingle, "Building the Foundation," 69.
40. CoAS annual reports (1924–25).
41. Patrick, "Final Report," 21. Patrick was consistent and unequivocal in his support of the ground-attack mission throughout his tenure as Air Service/Corps chief.
42. Maurer, *Aviation in the U.S. Army,* 80. The *Air Service Newsletter,* in this case, periodic issues for the years 1921 through 1923, thoroughly covered every aspect of unit aircraft acquisition and upgrades. In the case of the GA-1, it was simply too heavy for the "must use" Liberty engines: thus it was slow, required extended takeoff and landing distance, had a short range after a slow climb, and was not maneuverable.
43. This section partly extracted from Robert P. White, "The U.S. Army Air Service/Air Corps and the Aviation Industry, 1918–1926," unpublished Staff Research Paper, Air Staff History Office, 1997, based on the following sources: Bingle, "Building the Foundation"; Patrick, *United States in the Air;* John B. Rae, "Financial Problems of the American Aircraft Industry, 1906–1940," *Business History Review* (Spring 1965); Rutkowski, *Politics of Military Aviation Procurement;* Lampert Report; and Lassiter Board, Minutes.
44. Patrick office diary, 10 Oct. 1922.
45. Patrick, *United States in the Air,* 177. There were a total of forty-two separate investigations of the Air Service from 1918 through 1926. As Patrick complained, "one investigation after another instigated by different people with all sorts of motives and making different charges." In comparing Patrick's methods with the Department of Defense "privatization" effort now in vogue, Patrick was well ahead of his time in utilizing private industry to better meet the needs of the military establishment.
46. Patrick diary, entry for March 26, 1919.
47. Patrick, *United States in the Air,* 54–55.
48. Point Paper, "The Relation of Commercial Aeronautics to National Defense," 1923, Entry 228, Box 4, RG 18, NA.
49. Lt. Col. Fechet, CoAS, memorandum to Patrick, "Greater Support for Airways Section," 24 April 1922, Box 482, File 321.9, RG 18, NA.
50. CoAS annual report (1921), 43–44; Maurer, *Aviation in the U.S. Army,* 150–51.
51. See Pamphlet No. 10, *Outline of Organization and Functions of the Office of the Director or Air Service, U.S. Army,* 5 Nov. 1919 (also as File 167.41-1 at AFHRA); *Organization of Director of Air Service: Functions, Chief of Air Service,* 5 Oct. 1920, both in Box 482, File 321.9, RG 18, NA. Prior to this, the Airways Section was called the Civil Affairs Division, with duties described thusly: "Plans for the accomplishment of operative activities in connection with other Governmental Departments and commercial aviation, in which the use of the airplane is considered."

52. See, for example, Luther K. Ball, President, Manufacturers Aircraft Association, letter to Maj. W. H. Frank, Airways Section, 14 Dec. 1921, in File 321.9, Box 480, RG 18, NA.
53. Fechet to Patrick, "Greater Support for Airways Section."
54. Ibid., 2.
55. "Airways Section Progress Report," Dec. 1921 (Jan.–Dec. 1921), File 167.401-n-611, AFHRA.
56. Nick A. Komons, *Bonfires to Beacons: Federal Civil Aviation Policy under the Air Commerce Act, 1926–1938* (Washington, D.C.: U.S. Department of Transportation, GPO, 1978), 25.
57. Patrick speech, "Commercial Aviation and the National Defense," to Chicago Association of Commerce, 3 June 1925, 3, Entry 228, Box 1, RG 18, NA.
58. 1st Lt. Van Zandt, "The Development of Air Traffic in the Next Few Years," prepared as a press release, March 1923, Entry 228, Box 1, RG 18, NA; Bingle, "Building the Foundation," 109.
59. See, for example, "Press Release for Monday, December 29, 1924," Entry 228, Box 1, RG 18, NA.
60. Maurer, *Aviation in the U.S. Army,* 151.
61. CoAS annual reports (1923), 61–62, and (1924), 108.
62. "Relation of Commercial Aeronautics to National Defense," 5.
63. Maj. Gen. Mason M. Patrick, "Air Service and Air Transportation," *Engineers and Engineering* (Jan. 1925): 7.
64. See Patrick, memorandum to Davis, 10 March 1924, with attachment, "Government and the Aircraft Industry in America," Entry 228, Box 4, RG 18, NA.
65. Memorandum for CoAS, 2 April 1924, "Article on Commercial Aviation and the National Defense," Entry 228, Box 1, RG 18, NA.
66. Patrick, memorandum to Davis, "Government and the Aircraft Industry in America."
67. Ibid., 4.
68. Although Patrick did much to enhance commercial aviation efforts in the United States, by 1928 America still lagged far behind the Europeans. Writing in the same year, Patrick described with envy the regularly scheduled passenger routes between the European capitals. See *United States in the Air,* 127–29.
69. Patrick speech, "Air Transportation," given at Indianapolis, 29 May 1925, to the Society of Automotive Engineers, 2, Entry 228, Box 1, RG 18, NA.
70. MMP, letter to Elmer A. Johnson, 8 Oct. 1924, Entry 228, Box 5, RG 18, NA.
71. Maj. H. H. Arnold, "Commercial Possibilities of Aircraft in the West," *Journal of Electricity* (15 Oct. 1920): 364–65. It took some time for Benny Foulois, one of the earliest military air pioneers, to get in print on the issue, that being in 1929.
72. Patrick speech, "Air Transportation."
73. Ibid., 6.

74. Komons, *Bonfires to Beacons*, 22.
75. Ibid., 23.
76. Patrick, "Air Service and Air Transportation."
77. On 9 Feb. 1924, Patrick gave a radio broadcast from a Washington, D.C., station entitled, "The Round-the-World Flight by the Army Air Service," which details the consummate arrangements, and personal interest Patrick took with this project. See "Talk Broadcasted by Major General Mason M. Patrick," Entry 228, Box 1, RG 18, NA.
78. Ibid., 4.
79. See Maurer, *Aviation in the U.S. Army,* 186–88, and Map 7, for a brief synopsis and route of the world flight.
80. See "Many Notables Will Address Convention," *National Underwriter,* 30 Oct. 1924, 1; and "Notable Speakers for Convention in December: Air Service Chief Secured," *Insurance Field,* 31 Oct. 1924, 3, in Entry 228, Box 4, RG 18, NA.
81. "Notable Speakers for Convention in December," 3.
82. Elsbeth Freudenthal, *The Aviation Business: From Kitty Hawk to Wall Street* (New York: Vanguard Press, 1940), 75–76.
83. Ibid., 78–87.
84. Patrick office diaries, entry for 11 Oct. 1921.
85. Patrick, *United States in the Air,* 90–91.
86. President's Aircraft Board, *Hearings Before the President's Aircraft Board* (Morrow Board) (Washington, D.C.: GPO, 1925), 640–41.
87. Komons, *Bonfires to Beacons,* 22, 36–37; pamphlet, "National Aeronautic Association of the United States," 1922, Entry 228, Box 6, RG 18, NA; various speeches by Patrick, Entry 228, Boxes 1–2, RG 18, NA.

Chapter 10. The Fallout from the Lassiter Report and the Fall of Billy Mitchell

1. Patrick office diary; Patrick speech schedule/itinerary; invitations to speak, in Entry 228/229, Boxes 1–8, RG 18, NA.
2. See Maurer, *Aviation in the U.S. Army,* chap. 9.
3. *New York Times,* 1923–24. The editorial of 15 Aug. 1924 dealt with the U.S. Air Mail System.
4. Levine, *Mitchell,* 274; Davis, *Billy Mitchell Affair,* 197.
5. The four articles were: "American Aviation Radio Is Rotten," *Radio Broadcast* (Jan. 1923); "Aviation and Geology," *U.S. Air Service* (May 1923); "Recent Progress in Airplane Devices," *Review of Reviews* (June 1923); "Tiger Hunting in India," *National Geographic Magazine* (Nov. 1924).
6. Mitchell Papers, Mitchell's 201 File, Manuscript Division, LoC.
7. U.S. Air Force Oral History Interview: Lt. Gen. Ira C. Eaker, 23 March 1982, 28, File K239.0512-626, AFHRA; MS 33 (Hap Arnold/Murray Green Collection),

"Gen Eaker, 59–60, Mason Patrick & Airpower," Box 4, Card 2.23, para. 2–4, Arnold Papers.
8. Patrick, letter to Pershing, 27 Dec. 1922, and Marshall, letter to Patrick, 8 Jan. 1923, in Pershing Papers, MSS, Howard E. Coffin File, Special Collections, LoC; and Tate, *The Army and Its Air Corps,* 51 n. 29.
9. Secretary of War, annual reports (1921–23).
10. Sladen, letter to Patrick, 6 Feb. 1925, Entry 228, Box 7, RG 18, NA.
11. Tate, *The Army and Its Air Corps,* 30–31.
12. Ibid., 31.
13. Tate in *The Army and Its Air Corps* states: "The Air Service was the most expensive branch of the Army . . . and the Lassiter Program would have cost an estimated $90 million a year, more than a third of the Army budget." See 30–31 and n. 21. These figures were seemingly taken by Tate from a statement made by Gen. Fox Conner before the Morrow Board in Nov. 1925. It seems, though, that Conner's statement went beyond hyperbole: The total War Department appropriation for 1925 (military and nonmilitary items) was $348 million; of that amount approximately $15 million was for the Air Service, or 4.3 percent of the total War Department budget (and less than .70 percent of the total federal budget). As far as the Lassiter program is concerned, it was recommended that Congress make annual appropriations of $25 million per year for the Air Service. This is far from the $90-million figure cited by Conner and, subsequently, by Tate. Also see Rudolf Modley, ed., *Aviation Facts and Figures, 1945* (New York: McGraw-Hill, 1945), 54; Secretary of War, annual report (1925).
14. There is substantial documentation that Patrick believed he had made inroads with the General Staff regarding the need for adequate funding, the most obvious being the findings of the Lassiter Board. For Patrick's personal view on this issue see Patrick, letter to Maj. F. P. Reynolds, 29 Oct. 1923; Patrick, letter to P.R.G. Groves, April 3, 1924. Both in Entry 228, Boxes 5 and 2 (respectively), RG 18, NA. Also see Tate, *The Army and Its Air Corps,* 31, and Futrell, *Ideas, Concepts, and Doctrine,* 51–53.
15. Final Report, War Plans Division, File 2038, RG 165, NA; Tate, *The Army and Its Air Corps,* 33.
16. Maurer, *Aviation in the U.S. Army,* 73; Tate, *The Army and Its Air Corps,* 20, 31.
17. Denby, letter to Weeks, 18 Feb. 1924, Morrow Board Records.
18. Lampert Report, 1.
19. Patrick, *United States in the Air,* 89, 181.
20. Patrick, letter to Adjutant General, "Reorganization of Air Forces for National Defense" (hereafter Patrick reorganization letter), 19 Dec. 1924, File 321.9 a-1, Box 484, RG 18, NA; Tate, *The Army and Its Air Corps,* 36.
21. Patrick reorganization letter, 1, para. 4.
22. Brig. Gen. Drum (G-3), memorandum to ACS, War Plans Division, 7 Jan. 1925, War Plans Division File 888-22, RG 165, NA; Tate, *The Army and Its Air Corps,* 37.

23. Summerall's displeasure was made known to Patrick in a letter complaining of Mitchell's Hawaiian Island inspection report. Patrick answered Summerall in a masterly show of support for both individuals. All papers relating to Mitchell's Pacific inspection are in Mitchell Papers, Manuscript Division, LoC. Levine, in *Mitchell,* touches on this topic briefly (294–95); Davis, in *Billy Mitchell Affair,* devotes three chapters to it (159–92).
24. Maurer, *Aviation in the U.S. Army,* 127.
25. Mitchell Papers, Mitchell's 201 File, Manuscript Division, LoC.
26. Davis, *Billy Mitchell Affair,* 196–97.
27. Ibid., 197–98.
28. Lampert Report, 1689.
29. Ibid., 2265–76.
30. Ibid., 2758–59; Maurer, *Aviation in the U.S. Army,* 45.
31. Patrick, *United States in the Air,* 179–80; Davis, *Billy Mitchell Affair,* 201; Levine, *Mitchell,* 305; Maurer, *Aviation in the U.S. Army,* 45–46.
32. Futrell, *Ideas, Concepts, and Doctrine,* 46.
33. Maj. Gen. Duncan, letter to Patrick, 11 March 1925, Entry 228, Box 4, RG 18, NA.
34. Patrick, letter to Maj. Gen. Duncan, 14 March 1925, Entry 228, Box 4, RG 18, NA.
35. *National Aeronautic Association Review* 3, no. 6 (June 1925): 84.
36. Davis, *Billy Mitchell Affair,* 151.
37. CoAS annual reports (1923–25); Modley, ed., *Aviation Facts and Figures, 1945,* 54.
38. Davis, *Billy Mitchell Affair,* 218; Maurer, *Aviation in the U.S. Army,* 129.
39. ACS, *Organization of Military Aeronautics,* 69.
40. In addition to Morrow, the board consisted of Maj. Gen. James G. Harbord, retired; Rear Adm. Frank F. Fletcher, retired; Howard E. Coffin, an engineer and aeronautics expert of long repute; Sen. Hiram Bingham (R-Conn.), of the Committee on Military Affairs; Rep. Carl Vinson (D-Ga.) of the Committee on Naval Affairs; Rep. James S. Parker (R-N.Y.), chairman of the Committee on Interstate and Foreign Commerce; Judge Arthur C. Denison of the Sixth Circuit Court of Appeals; and William F. Durand of Stanford University, and engineer and National Committee for Aeronautics member. See Tate, *The Army and Its Air Corps,* 40–41. For a complete transcript of testimony by Air Service officers before the Morrow Board, see "U.S. President's Aircraft Board: Testimony of Gen. M. M. Patrick, Maj. W.J. Kilner, Maj. T. J. Lamphier and others before the Morrow Board Sept. 21-Oct. 13, 1925, Extract," File 248.211-61V, AFHRA.
41. Davis, *Billy Mitchell Affair,* 231.
42. "U.S. President's Aircraft Board: Testimony Extract," File 248.211-61V, AFHRA. Testimony also found in:"Verbatim Report of Morrow Commission Inquiry," in *Army and Navy Journal* (26 Sept. 1925): 8–10.
43. Futrell, *Ideas, Concepts, and Doctrine,* 1:45–46.

44. Morrow Board Report, 370–72, 377–79, 383–85, 389–92, 397–98, 403–4, 696, 719, 748, 763–64, 787, 808–9; Grumelli, "Trial of Faith," 131.
45. Gorrell, letter to Patrick, 9 Sept. 1925, Entry 228, Box 4, RG 18, NA. In his lead sentence, it is notable that Gorrell employed the word "publicity" when referring to Mitchell's statement.
46. Patrick, letter to Gorrell, 11 Sept. 1925, Entry 228, Box 4, RG 18, NA.
47. The entire transcript of the court-martial testimony can be found in "General Court Martial of Col. Wm Mitchell, Case #168771," vol. 27, Boxes 2914–2956, RG 94, NA. Patrick's testimony can be found on pages 2943–55. An excellent dissection and commentary of Patrick's testimony, along with others, can be found in Grumelli, "Trial of Faith." A description and critique of Patrick's testimony can be found on 235–41.
48. Grumelli, "Trial of Faith," 236; Ira C. Eaker, "Some Observations on Air Power," *Air Power and Warfare* (1979): 58.
49. Gen. Ira C. Eaker, interviewed by Lt. Col. Joe B. Green, Washington, D.C., 28 Jan. 1972, 28, on file at Army War College, Carlisle Barracks, Pa.
50. See Futrell, *Ideas, Concepts, and Doctrine,* 44–51, for Morrow Board overview.

Chapter 11. The Air Corps Act and Its Aftermath

1. Patrick, letter to Col. Frank E. Smith, 13 Jan. 1925, Entry 228, Box 7, RG 18, NA.
2. "Legislation and Its Effect on the Army Air Arm," chap. 1 of unpublished manuscript, "The United States Army Air Arm Between the Wars," 26, AFHSO.
3. For a detailed review of the convoluted legislative process during this period, see McClendon, *Autonomy of the Air Arm,* chap. 5.
4. "United States Army Air Arm Between the Wars," 26.
5. Copy of proposal found in File 321.9, Box 484, RG 18, NA. The documentation dealing with the give-and-take of the legislative process from January 1926 through to the passage of the Air Corps Act on 2 July is voluminous, and Patrick's role in his testimony before the various committees was extensive; but it was nothing that Patrick had not already said before. In each case, he called for eventual independence for the Air Force within a Department of Defense. In particular see File 321.9, Box 484, RG 18, NA.
6. Patrick reorganization letter, 1, para. 4.
7. Lampert Report, 521–32; Morrow Board, 1:72–73. Not only did Patrick speak of the concept, but Mitchell did as well when he referred to Patrick's proposal when in testimony before both the Lampert Committee and the Morrow Board. See Lampert Report, 2:1895–96.
8. Morrow Board, 2:1198.
9. Ibid., 2:72–73; Lampert Committee, 523.
10. Letter with attachment, 1st Endorsement, O.C.A.S. to Adjutant General, "Comment

on H.R. 8533," 3 Feb. 1926, File 321.9, Box 484, RG 18, NA. The General Staff (in the form of Maj. Gen. Fox Conner, G-4) critiqued the Patrick Bill and noted that it was as close to independence as one could get without actually achieving it. See Maj. Gen. Conner, memorandum to CoS, 1 Feb. 1926, File 580, RG 407, NA.
11. Morrow Board, 15–21.
12. 1st Endorsement, CoAS to Adjutant General, 13 Feb. 1926, Entry 032, Box 52, RG 18, NA; Bingle, "Building the Foundation," 124.
13. Letter, with attachment, Patrick to Adjutant General, 8 Jan. 1925, "Proposed Basis for 5 Year Expansion Program for Air Service to Comply with Recommendation of the President's Aircraft Board," File 321.9, Box 484, RG 18, NA.
14. Bingle, "Building the Foundation, 126–27.
15. War Department Training Regulation 440-15, *Fundamental Principles for the Employment of the Air Service,* 26 Jan. 1926, as cited in Futrell, *Ideas, Concepts, and Doctrine,* 50.
16. *Fundamental Principles,* 1.
17. Futrell, *Ideas, Concepts, and Doctrine,* 50.
18. With Billy Mitchell gone, according to Copp in *A Few Great Captains,* "Arnold was now the leader in a very one-sided battle that pitted a handful of undisciplined flying officers of junior rank against the massive bulk and power of the War Department and the Coolidge Administration." See p. 48.
19. Copp, *A Few Great Captains,* 48–49.
20. The complete investigation can be found in MS 17 Reports, Box 8, Series 4, Folder 4, USAFA/SC. When Arnold was initially confronted about the incident he denied his role but admitted all the following day. See "Statement of Major H. H. Arnold, Air Service," 17 Feb. 1926, Folder 4.
21. Copp, *A Few Great Captains,* 50: Tate, *The Army and Its Air Corps,* 46.
22. Levine, *Mitchell,* 373.
23. Dik Alan Daso, *Hap Arnold and the Evolution of American Airpower* (Washington, D.C.: Smithsonian Institution Press, 2000), 113.
24. Davis, *Billy Mitchell Affair,* 253; Daso, *Hap Arnold,* 112, 268 n. 60.
25. Arnold, *Global Mission* (draft in Murray Green Collection/USAFA), 44; Daso, *Hap Arnold,* 113; Davis, *Billy Mitchell Affair,* 255.
26. The Arnolds suggest that Hap was encouraged to engage in similar behavior—contacting local congressmen for support—while he was stationed in California. If this were indeed the case, and a court-martial of Arnold ensued, the defense would doubtless charge that Patrick was directly involved in these endeavors. No evidence, though, has surfaced to substantiate Patrick's participation in such matters.
27. *Hearings Before the Committee on Military Affairs, United States Senate on H.R. 10827, Congressional Record* 67, pt. 7, 6544 (Washington, D.C.: GPO, 1926), 1.
28. Bingle, "Building the Foundation," 129.

29. *Congressional Record* 67, pts. 9 and 10, 10674.
30. Also see Bingle, "Building the Foundation," 131–35.
31. ACS, *Organization of Military Aeronautics,* 77.
32. Ibid., 79, 119, 121.
33. For the most part, personnel authorizations were a shell game when it came to congressional authorizations and funding for same. The Air Service, which was authorized 1,516 officers and 16,000 enlisted men in the National Defense Act of 1920, rarely functioned with little more than half that strength. The Air Corps actually came into being with 919 officers and 8,725 enlisted in its ranks. The Air Corps Act authorized 1,517 officers and 16,000 enlisted. On the books, the Air Corps gained one officer authorization. See McClendon, *Autonomy of the Air Arm,* 59; Bingle, "Building the Foundation," 131; ACS, *Organization of Military Aeronautics,* 78–79; Maurer, *Aviation in the U.S. Army,* 9, 196. Also *Air Corps Act,* Public Law No. 446, 69th Congress, File 248.122-8, 6, AFHRA.
34. *Air Corps Act,* 1–2.
35. Ibid., 6; Bingle, "Building the Foundation," 133.
36. *Air Corps Act,* 6.
37. The Drum Board in the winter of 1933–34, followed by the Baker Board in 1934.
38. Futrell, *Ideas, Concepts, and Doctrine,* 1:51.
39. Of the seventeen total pages of the 1926 Air Corps Act, almost half were devoted to the changes in the equipment procurement procedure, a relatively minor portion of the overall legislation. This is indicative of the sensitivity of Congress to controlling the contract process.
40. Rutkowski, *Politics of Military Aviation Procurement,* 22–23.
41. Futrell, *Ideas, Concepts, and Doctrine,* 62.

Chapter 12. Conclusion

1. See Steven C. Call, "A People's Air Force: Air Power and American Popular Culture, 1945–1965," Ph.D. diss., Ohio State University, Columbus, 1997, chap. 2.
2. See also Trest, *Air Force Roles and Missions,* 49.
3. Patrick, "The Army Air Service," Army War College lecture, Carlisle Barracks, Pa., 9 Nov. 1925, on file at MHI. Also in Futrell, *Ideas, Concepts, and Doctrine,* 49, 53. This was a seminal presentation in Patrick's doctrinal evolution when he told his War College audience that he was going to persevere in his quest against the "reactionary plea" of those who continued to resist change, and noting the possibly decisive effects of air power on the national will.
4. Futrell, *Ideas, Concepts, and Doctrine,* 53.
5. John H. Smalls, letter to Patrick, 21 Dec. 1926, Entry 228, Box 7, RG 18, NA.
6. Patrick, letter to John H. Smalls, 8 Jan. 1927, Entry 228, Box 7, RG 18, NA, 1.
7. Ibid., 2–3.

8. The *Air Forces News Letter* 25, no. 1 (February 1942): 3, 4, 40; USAFA, MS 17, Box 19, Series 11, Folder 3; Myron Smith, "Mason Mathews Patrick," in *Dictionary of American Military Biography,* ed. Roger Spiller (Westport, Conn.: Greenwood Press, 1984), 2:826–29; *Army Register,* 1942, p. 1194; *Aviation* 23, no. 24 (12 December 1927): 1402–3; Bingle, "Building the Foundation," 139.
9. Lt. Gen. H. H. Arnold, letter to Bream Patrick, 3 February 1942, File 167.611-3, AFHRA.

Bibliography

Primary Sources

Government Archives

Army War College Curriculum Files, U.S. Army Military Institute, Carlisle Barracks, Carlisle, Pa.

Chief of the Air Corps. Annual reports (1926–27). U.S. Air Force Academy Library, Colorado Springs.

Chief of the Air Service. Annual reports (1920–25). U.S. Air Force Academy Library, Colorado Springs.

Dickman Board. File 167.404-6. USAF Historical Research Agency, Maxwell Air Force Base, Montgomery, Ala.

Lassiter Board. File 145.93-101/102. USAF Historical Research Agency, Maxwell Air Force Base, Montgomery, Ala.

Record Group Air 1, Public Record Office, London.

Secretary of War. Annual Report to the President (1916–27). Archives, Air Force History Support Office, Washington, D.C.

Personal Papers

Arnold, H. H. "Hap." MS 33, Murray Green Collection. Special Collections, U.S. Air Force Academy Library, Colorado Springs.

Bennett, Louis. Family Papers. R65E-2, Special Collections. West Virginia University Library, Morgantown.

Drum, Hugh A. Papers. U.S. Army Military Institute, Carlisle Barracks, Carlisle, Pa.

Eaker, Ira C. Papers. U.S. Army Military Institute, Carlisle Barracks, Carlisle, Pa.

Foulois Papers. MS 17, Manuscripts, Special Collections, U.S. Air Force Academy Library, Colorado Springs.
Jones Collection. MS 33. Special Collections, U.S. Air Force Academy Library, Colorado Springs.
Mitchell, William. Papers. Special Collections, Manuscript Division, Library of Congress, Washington, D.C.
Patrick, Maj. Gen. Mason M. Miscellaneous Papers. U.S. Army Aviation Museum, Fort Rucker Army Base, Dothan, Ala.
Patrick, Mason M. Papers, SMS 198, Manuscripts. Special Collections, U.S. Air Force Academy Library, Colorado Springs.
Patrick Family Records. Greenbrier County Historical Records, Lewisburg, W.V.
Pershing, John J. Papers. Manuscript Division, Library of Congress, Washington, D.C.

Published Congressional Records and Army Documents

President's Aircraft Board. *Hearings Before the President's Aircraft Board* (Morrow Board). Washington, D.C.: GPO, 1925.
———. *Report of the President's Aircraft Board*. Washington, D.C.: U.S. GPO, 1925.
United States Army in the World War, 1917–1919, Reports of the Commander-in-Chief, Staff Sections and Services. Washington, D.C.: Center of Military History, U.S. Army, 1991.
U.S. Congress. House. Committee on Appropriations. *Hearings on War Department Appropriations Bill*. 67th–69th Congresses, 1923–28.
U.S. Congress. House. Committee on Military Affairs. *Hearings of Defense and Unification of Air Service*. 69th Cong., 1st sess., 1926.
———. Frear Subcommittee Report. 66th Cong., 2nd sess., 1919, H. Rept. 637.
———. *Hearings on H.R. 10147 and H.R. 12285 to Create a Department of Aeronautics, Defining the Powers and Duties of the Secretary Thereof, Providing for the Organization, Disposition, and Administration of a United Air Force, and Providing for the Development of Civil and Commercial Aviation, the Regulation of Air Navigation, and for other Purposes*. 68th Cong., 2nd sess., 1925.
———. Inquiry into Operations of the United States Air Service. *Hearings Before the Select Committee of Inquiry into Operations of the United States Air Service*. 68th Cong., 2nd sess., 1924.
U.S. Congress. House. Select Committee of Inquiry into Operations of the United States Air Services. *Hearings on Matters Relating to the Operations of the United States Air Services*. 68th Cong., 1st sess., 1925.
———. *Report of the Select Committee of Inquiry into Matters Relating to the Operations of the United States Air Services*. House Report No. 1653. 68th Cong., 2nd sess., 1925.
U.S. Congress. Senate. Committee on Military Affairs. *Hearings on H.R. 10827: An Act to Provide More Effectively for the National Defense Increasing the Efficiency of the*

Air Corps of the Army of the United States, and for Other Purposes. 69th Cong., 1st sess., 1926.

———. Hearings on S. 2614: *A Bill to Increase the Efficiency of the Air Service of the United States Army.* 69th Cong., 1st sess., 1926.

U.S. War Department, *Field Service Regulations, United States Army, 1923.* Washington, D.C.: GPO, 1924.

Official Studies

Assistant Chief of Staff, Intelligence, Historical Division. *Organization of Military Aeronautics, 1907–1935.* Army Air Forces Historical Studies no. 25. Washington, D.C.: Army Air Forces Historical Office, 1944.

———. *Material Research and Development in the Army Air Arm, 1914–1945.* Army Air Forces Historical Studies no. 50. Army Air Forces Historical Office, 1945.

Layman, Martha. *Legislation Relating to the Air Corps Personnel and Training Programs, 1907–1939.* Army Air Forces Historical Studies no. 39. Army Air Forces Historical Office, 1945.

Parker, John D. *The Early Development of United States Air Doctrine.* Professional Study no. 6024. Maxwell Air Force Base, Ala.: Air War College, 1976.

U.S. Air Force. *The Development of Air Doctrine in the Army Air Arm, 1917–1941.* USAF Historical Studies no. 89. Maxwell Air Force Base, Ala.: Historical Division, Air University, 1955.

———. *U.S. Air Service Victory Credits, World War I.* USAF Historical Study no. 133. Maxwell Air Force Base, Ala.: Historical Research Division, Air University, 1969.

Secondary Sources

Books

Adas, Michael. *Machines as the Measure of Men: Science, Technology, and Ideologies of Western Dominance.* Ithaca N.Y.: Cornell University Press, 1989.

Arnold, H. H. "Hap." *Global Mission.* New York: Harper, 1949.

Bidwell, Shelford. *Fire-Power: British Army Weapons and Theories of War 1904–1945.* Boston: Allen and Unwin, 1985.

Bilstein, Roger. *The American Aerospace Industry, from Workshop to Global Enterprise.* New York: Twayne Publishers, 1996.

Braim, Paul F. *The Test of Battle: The American Expeditionary Force in the Meuse-Argonne Campaign.* Newark, N.J.: White Mane Publishers, 1983.

Burlingame, Roger. *General Billy Mitchell.* New York: Greenwood Press, 1972.

Chatfield, Charles. *The American Peace Movement: Ideals and Activism.* New York: Twayne Publishers, 1991.

Cameron, Rebecca Hancock. *Training to Fly: Military Flight Training, 1907–1945.* Washington, D.C.: Air Force History and Museums Program, 1999.

Coffman, Edward. *The War to End All Wars*. New York: Oxford University Press, 1968.
Cooke, James J. *Pershing and His Generals, Command and Staff in the A.E.F.* Westport, Conn.: Praeger, 1997.
———. *The U.S. Air Service in the Great War, 1917–1919*. Westport, Conn.: Praeger, 1998.
Copp, DeWitt S. *A Few Great Captains*. McLean, Va.: EPM Publications, 1980.
Craven, Wesley F., and James L. Cate, eds. *Plans and Early Operations, January 1939–August 1942*. Vol. 1 of *The Army Air Forces in World War II*. Chicago: University of Chicago Press, 1948.
Crowell, Benedict, and Robert F. Wilson. *Demobilization*. Vol. 4 of *How America Went to War*. New Haven: Yale University Press, 1921.
Cullum, George W. *Biographical Register of the Officers and Candidates of the U.S. Military Academy at West Point, New York,* Vol. VI-A. Saginaw, Mich.: Seeman and Peters, 1920.
———. *Biographical Register of the Officers and Graduates of the U.S. Military Academy at West Point, New York, since Its Establishment in 1802, to 1890*. Third ed. Vol. 3. New York: R. R. Donnelley and Sons, 1891.
Daso, Dik Alan. *Hap Arnold and the Evolution of American Airpower*. Washington, D.C.: Smithsonian Institution Press, 2000.
Davis, Burke. *The Billy Mitchell Affair*. New York: Random House, 1967.
Finney, Robert T. *History of the Air Corps Tactical School, 1920–1940*. Washington, D.C.: Office of Air Force History, 1994.
Foulois, Benjamin D., and Carroll V. Glines. *From the Wright Brothers to the Astronauts: The Memoirs of Major General Benjamin D. Foulois*. New York: McGraw-Hill, 1968.
Fredette, Raymond H. *The Sky on Fire: The First Battle of Britain, 1917–1918*. Washington, D.C.: Smithsonian Institution Press, 1991.
Freudenthal, Elsbeth. *The Aviation Business: From Kitty Hawk to Wall Street*. New York: Vanguard Press, 1940.
Futrell, Robert Frank. *Ideas, Concepts, and Doctrine: Basic Thinking in the United States Air Force 1907–1960*. Maxwell AFB, Ala.: Air University Press, 1989.
Gibbs-Smith, Charles H. *A History of Flying*. London: Batsford, 1953.
Goldberg, Alfred, ed. *A History of the United States Air Force, 1907–1957*. Princeton, N.J.: D. Van Nostrand, 1957.
Hallion, Richard P. *Strike from the Sky: The History of Battlefield Air Attack, 1911–1945*. Washington, D.C.: Smithsonian Institution Press, 1989.
———. *Rise of the Fighter Aircraft*. Baltimore: Nautical and Aviation Publishing Company of America, 1988.
Hennessy, Juliette. *The United States Army Air Arm, April 1861 to April 1917*. Washington, D.C.: Office of Air Force History, 1985.

Holley, I. B., Jr. "An Enduring Challenge: The Problem of Air Force Doctrine." *The Harmon Memorial Lectures in Military History,* no. 16. Colorado Springs: U.S. Air Force Academy, 1974.

———. *Ideas and Weapons: Exploitation of the Aerial Weapon by the United States During World War I.* 1953. New imprint. Washington, D.C.: Office of Air Force History, 1983.

Hudson, James J. *Hostile Skies.* Syracuse, N.Y.: Syracuse University Press, 1968.

Hurley, Alfred F. *Billy Mitchell: Crusader for Air Power.* Bloomington: Indiana University Press, 1975.

Kennett, Lee. *The First Air War, 1914–1918.* New York: Free Press, 1991.

Komons, Nick A. *Bonfires to Beacons: Federal Civil Aviation Policy under the Air Commerce Act, 1926–1938.* Washington, D.C.: U.S. Department of Transportation, GPO, 1978.

Lahm, Frank P. *The World War I Diary of Colonel Frank P. Lahm, Air Service, A.E.F.* Maxwell AFB, Ala.: Historical Research Division, Air University, 1970.

Levine, Isaac Don. *Mitchell, Pioneer of Air Power.* New York: Duell, Sloan and Pearce, 1958.

MacIsaac, David. "The Air Force." In *Encyclopedia of the American Military,* ed. John E. Jessup and Louise B. Katz. New York: Scribner's, 1994.

Maurer, Maurer. *Aviation in the U.S. Army, 1919–1939.* Washington, D.C.: Office of Air Force History, 1987.

Maurer, Maurer, ed. *The U.S. Air Service in World War I* (Gorrell). Washington, D.C.: Office of Air Force History, 1978.

McClendon, R. Earl. *Autonomy of the Air Arm.* Washington, D.C.: Air Force History and Museums Program, 1996.

Modley, Rudolf, ed. *Aviation Facts and Figures, 1945.* New York: McGraw-Hill, 1945.

Morrow, John H. *German Air Power in World War I.* Lincoln: University of Nebraska Press, 1982.

———. *The Great War in the Air: Military Aviation from 1909 to 1921.* Washington, D.C.: Smithsonian Press, 1993.

Mitchell, William. *Memoirs of World War I: From Start to Finish of Our Greatest War.* Reprint. Westport, Conn.: Greenwood Press, 1975.

———. *Our Air Force: The Keystone of National Defense.* New York: E. P Dutton, 1921.

———. *Skyways: A Book on Modern Aeronautics.* Philadelphia: Lippincott, 1930.

———. *Winged Defense.* New York: Putnam's and Sons, 1925.

Murray, Williamson, and Allan R. Millett, eds. *Military Innovation in the Interwar Period.* Cambridge: Cambridge University Press, 1996.

Nalty, Bernard C., ed. *Winged Shield, Winged Sword: A History of the United States Air Force.* 2 vols. Washington, D.C.: Air Force History and Museums Program, 1997.

Patrick, Mason M. *Military Aircraft and Their Use in Warfare*. Philadelphia: The Franklin Institute, 1924.

———. *The United States in the Air*. Garden City, N.J.: Doubleday, Doran, 1928.

Pershing, John J. *My Experiences in the World War*. 2 vols. New York: Frederick A. Stokes, 1931.

Rutkowski, Edwin H. *The Politics of Military Aviation Procurement, 1926–1934*. Columbus: Ohio State University Press, 1966.

Schivelbusch, Wolfgang. *The Railroad Journey: The Industrialization of Time and Space in the Nineteenth Century*. Berkeley: University of California Press, 1986.

Sherry, Michael S. *The Rise of American Airpower: The Creation of Armageddon*. New Haven, Conn.: Yale University Press, 1987.

Shiner, John F. *Foulois and the U.S. Army Air Corps, 1931–1935*. Washington, D.C.: Office of Air Force History, 1983.

Smythe, Donald. *Pershing: General of the Armies*. Bloomington: Indiana University Press, 1986.

Tate, James Phillip. *The Army and Its Air Corps, 1919–1941*. Maxwell Air Force Base, Ala.: Air University Press, 1998.

Thayer, Lucien. *America's First Eagles: The Official History of the U.S. Air Service, A.E.F., 1917–1918*. San Jose, Calif.: Bender Publishing; Mesa, Ariz.: Champlin Fighter Museum Press, 1983.

Toulmin, H. A., Jr. *Air Service, A.E.F., 1918*. New York: D. Van Nostrand, 1927.

Trest, Warren A. *Air Force Roles and Missions: A History*. Washington, D.C.: Air Force History and Museums Program, 1998.

Trimble, William F. *Admiral William A. Moffett: Architect of Naval Aviation*. Washington, D.C.: Smithsonian Institution Press, 1994.

Weigley, Russell F. *History of the United States Army*. New York: Macmillan, 1967.

Underwood, Jeffrey S. *The Wings of Democracy: The Influence of Air Power on the Roosevelt Administration, 1933–1941*. College Station: Texas A&M Press, 1991.

Articles

Arnold, Major H. H. "Commercial Possibilities of Aircraft in the West." *Journal of Electricity* (Oct. 15, 1920): 364–65.

"Bombs vs. Boats." *New York Times,* Sept. 14, 1921, 1.

Chamberlain, Sen. George E. "Military Aeronautics: No General Staff Control of Army Air Service." *U.S. Air Service* 1, no. 4 (May 1919): 13–15.

Dargue, Major H. A. "Training a Major General to Fly." *U.S. Air Service* (September 1923): 18–22.

Eaker, Ira J. "Some Observations on Air Power." *Air Power and Warfare* (1979): 58.

"Generals Hines and Patrick Remembered." *Beckley* (W.V.) *Post-Herald,* September 1, 1978, 3.

"General Mason Patrick, Air Corps Chief Honored." *Lewisburg* (W.V.) *Gazette,* July 2, 1978, 2.
Gross, Charles J. "George Owen Squier and the Origins of American Military Aviation." *Journal of Military History* (July 1990): 285–92.
"Leadership of the Air Service." *Washington Star,* August 13, 1919, 1.
"Maj. Gen. Mason M. Patrick: Chief of the U.S. Army Air Service." *Time,* July 9, 1923.
"Many Notables Will Address Convention." *The National Underwriter,* October 30, 1924, 3.
Menoher, Maj. Gen. Charles T. "The Future of the Air Service." *U.S. Air Service* 1, no. 3 (April 1919): 10–11.
Mitchell, Brig. Gen. William. "Air Leadership." *U.S. Air Service* 1, no. 4 (May 1919): 16–17.
"Notable Speakers for Convention in December: Air Service Chief Secured." *The Insurance Field,* October 31, 1924, 3.
Patrick, Maj. Gen. Mason M. "Air Service and Air Transportation." *Engineers and Engineering* (January 1925): 31–32.
———. "Aviation in the United States during 1923." *L'Air,* November 1923, 8–9.
———. "The Development of Air Traffic in the Next Few Years." *New York Times,* 13 March 1923, 2.
———. "Recommendations for Improving the Air Service." *U.S. Air Service* (February 1923): 15–16.
———. "The Use of Aircraft in War." *American Unity Forward,* October 1922, 12–24.
Smith, Myron J., Jr. "Mountaineer Boss of the Eagles." *Lewisburg Gazette,* July 2, 1978, 3.
"Verbatim Report of Morrow Commission Inquiry." *Army and Navy Journal,* September 26, 1925.
Wainwright, J. Mayhew. "The Situation in the Air Service." *U.S. Air Service* (April 1923): 19.

Unpublished Materials

Bingle, Bruce A. "Building the Foundation: Major General Mason Patrick and the Army Air Arm, 1921–1927." M.A. thesis, Ohio State University, 1981.
Call, Steven C. "A People's Air Force: Air Power and American Popular Culture, 1945–1965." Diss., Ohio State University, 1998.
Flugel, Raymond R. "United States Air Power Doctrine: A Study of the Influence of William Mitchell and Giulio Douhet at the Air Corps Tactical School, 1921–1935." Ph.D. diss., University of Oklahoma, 1965.
Grumelli, Michael L. "Trial of Faith: The Dissent and Court-Martial of Billy Mitchell." Ph.D. diss., Rutgers University, New Brunswick, N.J., 1991.
Hallion, Richard P. "World War I Aviation." Air Power History Course. Air Staff History Office. Pentagon, Washington, D.C., 10 February 1997.

"Legislation and Its Effect on the Army Air Arm." Chapter 1 of "The United States Army Air Arm Between the Wars." Washington, D.C.: Air Force History Support Office, n.d.

"Materiel Organization." Chapter 7 of "Early History of the Air Service." Air Force History Support Office, Washington, D.C., n.d.

Miller, Roger G. "Keep 'Em Flying: A History of Air Force Logistics from the Mexican Border to the Persian Gulf." Air Force History Support Office, Washington, D.C., 1997.

Ransom, Harry H. "The Air Corps Act of 1926: A Study in the Legislative Process." Ph.D. diss., Princeton University, 1953.

White, Robert P. "The U.S. Army Air Service/Air Corps and the Aviation Industry, 1918–1926." Staff Paper, Air Staff History Office, 1997.

Index

Air Bill (H. R. 7916), 123
Air Commerce Act, 136
Air Corps Act (1926), 4, 6, 71, 72, 122–31, 133–35
Air Corps Materiel Division, Wright Field, 93–102, 106, 128
Air Corps Training Center, 128
Aircraft (Production) Board, 27
Air Mail Act (Kelly Bill), 108
Air Mail System (U.S.), 104
Air Service, Airways Section, 103; appropriations, 4, 12, 26, 68, 87, 100, 112–13, 129; coastal defense, 91–93; demobilization post–WW I, 44–46; discipline, 6, 61–62; doctrine, 2, 6–7, 10–11, 24–25, 82; personnel, 3, 4, 63, 68, 79; pilot fatalities (post–WW I), 95
Air Service Act (1920), 68
Air Service Board, 38
American Aviation Mission, 38, 39
American Expeditionary Force (AEF), Air Service, aviators and discipline, 18–20, 33, 47; in WW I, 11–35; leadership, 3, 18–20, 22; mission and doctrine, 23–25, 33–34; planes and personnel, 3, 21, 25–28, 30, 34; WW I casualties, 23
American Legion, 108
Ames, Carter, 108
Aquitania, 41, 42
Army Reorganization Act, 50–52, 67

Army Reorganization (Baker/March) bill, 47, 48
Army War College, 66, 89–90
Arnold, Eleanor ("Bee"), 127
Arnold, Henry Harley ("Hap"), 24, 57, 81, 92, 106, 110, 126–28, 134–5, 137
Around the World Flight, 107, 110
Aisne-Marne campaign, 133
Aeronautics Commission of the Supreme War Council, 37
aviation, American, early character, 2, 8–11; military air doctrine, 10, 24; funding, 2, 8, 9–10, 12, 26, 28, 34; technology, 4, 9, 11
aviation, European, 2, 8–9, 105

Baker, Newton D., 38, 39, 45, 51
Bane, Thurman H., 94
Bell, George, Jr., 47
Bennett, Louis, Jr., 40–41
Bennett, Louis, Sr., 40
Bennett, Sallie Maxwell, 40–41, 58
Blaine, Gerald, 29
Blatchford, R. M., 18
Bliss, Tasker H., 30–31, 36
Board of Ordnance and Fortification, 8
Bolling Commission, 12, 34
Borah, William E., 54
Budget and Accounting Act of 1921, 67
Bureau of Aircraft Production, 13, 27
Bureau of the Budget, 63, 67–68, 120–21
Bureau of Naval Aeronautics, 53

Index

Camp Humphreys, 56
Chateau-Thierry, 22
civil aviation, 102–9, 135
Coffin, Howard E., 39,
Coolidge, Calvin, 5, 67, 70, 87, 99, 105, 116, 119, 120, 122–23, 130–31, 136
Conner, Fox, 112–13
Craig, Malin, 21
Crowell, Benedict, 38, 39, 41
Crowell Commission, 49
Curry, John F., 98
Curry Bill, 11, 47, 48, 77

Davis, Dwight F., 83, 88, 97, 98, 102, 105–6, 126
Davison, F. Trubee, 132
Dargue, Herbert A., 77, 85, 126, 134–35
Denby, Edwin, 86, 114
Department of Aeronautics, 39
Dickman, Joseph, 38, 42
Dickman Board, 38, 47, 49, 64
Doolittle, James H. ("Jimmy"), 107
Drum, Hugh A., 22, 33, 56, 77, 112, 115, 126
Duncan, George B., 117

Eaker, Ira, 74, 111, 127
Ely, Hanson E., 89

Fechet, James E., 65, 74, 104, 117, 126
Fiske, Harold, 42
Foulois, Benjamin ("Benny") D., Air Service, AEF, 2, 3, 12, 13, 19–22, 29, 33, 105; Air Service Board, 38; early military aviation regulations, 10, 12; pre–WW I, 10, 24; Punitive Expedition, 27; rationale for Air Service independence, 49; relationship with Mitchell, 12, 51
Foulois Board, 47, 64
Frank, William H., 73
Frear Committee, 42, 43
"Fundamental Conceptions," 72–83, 90, 134
Fundamental Doctrine of the Air Service, 65

General Headquarters (GHQ) Air Force, 66, 76, 126
General Order No. 4, 92
General Order No. 20, 126
General Service Schools, 90–91
General Staff, 27, 34
GHQ Reserve, 77, 79
Gorrell, Edgar S., 33, 41, 119

Harbord, James G., 18, 59, 60–61, 69, 73
Harding, Warren G., 5, 59, 67, 87, 103, 105, 107
Hart, Liddell, 135
Harvard Engineering Society, 89
Hay, James, 10
Heintzelman, Stuart, 76
Hickam, Horace M., 119
Hill, William, 126
Hines, John L., 16
Hodges, H. F., 56
Hoover, Herbert, 107, 136–37
House Committee on Military Affairs, 48, 52, 123, 124, 125, 126, 127, 128

Independent Air Force (British), 28, 31
Inter-Allied Aviation Committee, 30
Inter-Allied Bombing Force, 30–31
Ireland, R. W., 99

James, W. Frank, 124
James Bill, 125

Kenly, William L., 13
Kernan, Francis J., 61
Kilner, Walter G., 119
Kilpatrick, Thomas, 15
Knapp, Silas, 37

Lahm, Frank P., 33, 77
Lampert Committee, 115–16, 118, 119, 122, 123, 124, 126, 127
Lampert Report, 119, 120, 121, 122, 135
Langley, Samuel P., 8
Lansdowne, Zachary, 118–19
Lassiter, William, 76, 77
Lassiter Board (Committee/Report), 69–84, 97, 98, 111–12, 114, 125, 130
Lassiter Program (War Department Major Project Number 4), 112–14
Lejeune, John A., 123
Liggett, Hunter, 21, 33, 42

Markham, Charles, 108
Marshall, George C., 112
Manchu Law, 10
March, Peyton C., 13, 21
McAndrew, James W., 19, 20, 21
McIntosh, L. W., 98, 99
Menoher, Charles T., clashes with Mitchell, 3, 4, 53, 59; Air Service commander, 25, 44, 42, 46, 50, 53–56, 60, 65, 74, 103, 104, 133, 134,

135; philosophy on Air Service independence, 45; resigns as Air Service chief, 55, 58
Menoher Board, 46, 47, 49, 77
Meuse-Argonne campaign, 22, 47, 133
Miller, Elizabeth (Mitchell's second wife), 87, 111
Milling, Thomas DeWitt, 119
Mitchell, Caroline Stoddard (Mitchell's first wife), 73
Mitchell, Ruth (Mitchell's sister), 12
Mitchell, William ("Billy"), aviation vision, 4; bombing of USS *New Jersey* and USS *Virginia*, 86–88; case for Air Service independence, 52; Calvin Collidge, 87; civil aviation control, 108, 135; clashes with Menoher, 53; court martial, 5, 120–21; in Cuba, 17; divorce, 73; fundamental conceptions, 83; historically overshadows Patrick, 1, 5; in WW I, 3, 11, 12, 13, 19–24, 27, 31–35; Lampert Committee, 116–17; not reappointed as Air Service deputy, 117; opposes separate Army aviation branch, 10; *Ostfriesland*, 53; press conference (5 September 1925), 119–20; promoted to brigadier general, 23; publicist, 5, 53, 59; published works, 111; rationale for Air Service independence, 49; relationship with AEF senior officers, 19–23, 48; relationship with U.S. Navy, 49, 50, 52, 54, 59; resigns from U.S. Army, 4, 5, 120; returns from Europe after WW I, 4, 42; sent to Europe (1921–22), 62
Model Airway System, 104–5
Moffett, William A., 53, 85
Morin, John, 123
Morin Bill, 124
Morrow, Dwight W., 119
Morrow Board (President's Aircraft Board), 119, 120, 122, 124, 125, 128, 129
Morrow Report, 121–22, 123

National Advisory Committee for Aeronautics (NACA), 12, 108, 117
National Aeronautic Association (NAA), 108, 116
New, Harry S., 46
New Bill, 47, 48, 77
Night Bombardment Division (AEF, Air Service), 28
Nolan, Dennis E., 30, 33, 42

Ostfriesland, 53, 54, 56, 69, 86, 89, 111, 118
Overman Act, 13

Patrick, Alfred S. (father), 15, 16
Patrick, Bream C. (son), 41, 137
Patrick, Grace Cooley (wife), 19, 41, 137
Patrick, Marsena R., 17
Patrick, Mason M., air force mobility, 91; Air Service discipline, 5, 6; and Benedict Crowell, 38; as engineer, 3, 16–18, 26, 89; bombing of USS *New Jersey* and USS *Virginia*, 86–88; centralized command of air forces, 76; changes view of military aviation role, 6, 66–67, 74, 132, 133–34; changing General Staff opinion of Air Service, 132; civil aviation, 102–9; commander of Air Service, AEF, 14, 18–43; in Cuba, 17; death, 137; doctrine, roles and missions, 4, 7, 11, 24–25, 61, 64–69, 86, 90–91, 134; first head of Air Service to emphasize offensive role of aircraft, 66, 73; first statement of airpower doctrine accepted by War Department, 64–69, 71, 83–84, 134; formative years, 15–16; Hap Arnold, 81, 92, 126–28, 134; Howard E. Coffin, 39; insurance industry, 108; investigations of Air Service, 102; Lampert Committee testimony, 119; learns to fly, 1, 84–86; mandatory retirement, 1, 130; McCook Field Engineering Division, 94–101; Mitchell's court martial, 120–21; post–WW I Air Service independence, 39, 115; in post–WWI Europe, 3, 36–41; promoted to brigadier general, 17; proposes creation of Air Corps, 115–16; reform of contracting system, 101–2; relationship with Mitchell, 2–5, 22–23, 28, 32–35, 56–57, 60–61, 73–74, 75, 111, 113, 117; relationship with Pershing, 3, 16, 18; returns from Europe after WW I, 42; returns to head Air Service (1921), 3–4, 34, 56–57; 59–61; service in AEF, 17–41; strategic bombing, 28–32; Versailles Peace Treaty terms, 37; West Point, 3, 15–16
Patrick, Virginia M., (mother), 15
Patrick Bill, 124, 126, 135
Pershing, Gen. John J., AEF commander, 2, 3, 5, 12, 13, 14, 18, 27, 29, 32–35; chooses Patrick to head Air Service, AEF, 14, 18–19, 132, 136; control of post–WW I Air Service, 5, 18–22, 26, 29; Final Report, 100; Punitive Expedition, 17; retirement, 114;

Pershing, Gen. John J., AEF commander (*continued*), returns from Europe after WW I, 43, 47; role in Patrick's return to head Air Service, 56–57, 135; strategic bombing, 30; views on Air Service, 48; West Point, 2, 16
Punitive Expedition (Mexico), 2, 9, 17, 27, 47

Rickenbacker, Eddie, 56
Robison, Samuel S., 86
Roosevelt, Theodore, 86
Royal Air Force, 31
Royal Flying Corps, 28, 32
Ryan, John D., 13, 14, 35

Salmond, J. M., 32
Scaife, Ronald, 102
Shenandoah, 118
Sherman, William C., 65, 68, 112
St. Mihiel, 22, 23, 27, 32, 64, 133
Signal Corps, U.S. (Aviation Section), 9, 10, 12, 13, 24–25, 27
Six Army Plan, 78
Sladen, Fred W., 112
Smith, H. A., 42
Smuts Commission, 31
Spaatz, Carl ("Tooey"), 127
Squier, George O., 12, 13
strategic bombing, 3, 28–32, 67–68, 76
Summerall, Charles P., 116
Supreme War Council, 30, 31, 36

surplus aircraft, 42–43, 63, 70, 79, 95, 130
Sykes, Frederick H., 30–31

Training Regulation 10–5, 82, 83
Training Regulation 440–15, 68, 82, 83, 91, 114, 126
Trenchard, Hugh, 28–34

The United States in the Air, 136

Van Zandt, James E., 105
Volandt, Roger, 67, 90, 97

Wadsworth, James, 126
Wadsworth Bill, 128
Wainwright, J. Mayhew, 75, 79, 83, 124
Wainwright Bill, 75, 124, 126
War Department General Order No. 4, 92
Washington Conference for the Limitation of Naval Armaments (Washington Naval Treaty), 54, 62, 86
Wells, Briant H., 76
Weeks, John W., 53, 55, 62, 65, 69, 75, 76, 79, 80, 81, 83, 97, 98, 111, 114, 116, 117, 127
Weir, Lord, 28, 31
Westover, Oscar, 46, 51
West Point, 3, 15, 16, 17, 76, 112, 136
Wilbur, Curtis, 114
Wilson, Woodrow, 13, 27, 45
Wright brothers, 8, 12, 33